T3-BNY-766

new & tigers OLD ELEPHANTS

Scott B. MacDonald

Jane E. Hughes

David Leith Crum

new tigers & OLD ELEPHANTS

the development game in the 1990s and beyond

Transaction Publishers

New Brunswick (U.S.A.) and London (U.K.)

This book is printed on acid-free paper that meets the American National Standard for Permanence of Paper for Printed Library Materials.

Library of Congress Catalog Number: 95-18142
ISBN: 1-56000-204-2
Printed in the United States of America

Library of Congress Cataloging-in-Publication Data

MacDonald, Scott B.
New tigers and old elephants : the development game in the 1990s and beyond / Scott B. MacDonald, Jane E. Hughes, David Leith Crum.
p. cm.
ISBN 1-56000-204-2 (alk. paper)
1. Developing countries—Economic policy. 2. Developing countries—Economic conditions. 3. Economic stabilization—Developing countries. I. Hughes, Jane. II. Crum, David L. III. Title.
HC59.7.M21116 1995
338.9'009172'4—dc20 95-18142
CIP

Contents

Acknowledgments

The process of writing this book was the collective work of the three authors, who take complete responsibility for the accuracy of the facts. As such, the views expressed in this volume are solely those of the authors and do not necessarily represent those of CS First Boston, Brandeis University, or NationsBank. At the same time, a number of people were kind enough with their time in enhancing the quality of the book's scholarship.

Scott B. MacDonald

I wish to extend my appreciation to Professor Albert L. Gastmann, formerly of Trinity College's Political Science Department; longtime friend Leon S. Tarrant, from the Office of the Comptroller of the Currency; and Dr. Norman Bailey for their comments on earlier drafts. Special thanks are also extended to Professor Ranbir Vohra, Chair of the Political Science Department at Trinity College, and J. Curtis Shambaugh of CS First Boston for their comments on the China chapter, and Uwe Bott at Moody's Investor Service for his views on international capital markets. Above all else, I appreciate the patience of my wife Kateri. Together with my son Alistair, they greatly enrich my life.

Jane H. Hughes

I wish to thank Scott and Dave. They have been ideal partners, especially in the final push. Most important, they never lost their senses of humor—even when things did not seem too funny.

There are a few other people who have inspired and supported me along the way. Bill Coplin of Political Risk Services, with whom I have worked for an incredible thirteen years. He never fails to challenge, interest, stimulate (and sometimes amuse) me. Anne Carter of Brandeis Economics Department, one of the true role models of my life. My father, who fights his way back from things that would knock most people down. My mother, who defies description! My husband, the most loving and giving person I have ever known. Alex, Zack, and Caroline, for being there.

David L. Crum

I want to thank all those who provided insight and assistance in the process of pulling together this volume, including my co-authors Jane and Scott, and my wife, Indira.

Preface

As we approach the turn of the twentieth century, the world in which we conduct our affairs is changing rapidly. Watching the fall of communism, the 1991 Gulf War, and the struggles to find a post-cold war international system is both fascinating and bewildering. For the developing world—stretching from the Andes Mountains and Caribbean Sea through Africa to the shores of the Indian and Pacific oceans—the so-called new world order remains undefined and the parameters hazy. The notion of this new international structure also carries a sense of unease, which opinion makers have readily noted. According to *The Wall Street Journal,* "The New World Order is beginning to look like a bad place to be a poor nation."[1] Sir Peter Leslie, chairman of the Commonwealth Development Corporation, also noted that in the next decade developing countries face a "period of neglect"[2] from the industrialized north, to which Ian Davidson adds, "There is a new world, with the disintegration of the Soviet Union; but there is no order, in the sense of a coherent and structured framework for future international policy...the defining characteristics of the new world will probably be disorder and unpredictability."[3]

Amongst all this uncertainty, at least one thing appears clear—many of the advantages that propelled countries forward in the cold war decades that followed World War II will no longer apply. Cheap labor, abundant raw materials, and strategic locations will not matter as much in a world no longer obsessed by superpower rivalries and increasingly driven by economic empowerment related to the harnessing of new technological breakthroughs. Ideology is out and pragmatism is in. Countries that have long benefited from their proximity to geopolitical hot spots (the Persian Gulf conflagration notwithstanding) will matter less. Even worse, in a world dominated by three emerging trade blocs centered on North America, Western Europe, and East Asia, those locked out of the global trading patterns will find it more difficult to export their way to prosperity.

The development challenges facing African, Asian, and Latin American countries are immense, but not insurmountable. For these developing nations, we assume that the decades of the 1990s and beyond will be characterized by an upsurge in ethnic and religious discord; by the waning, or even irrelevance, of many of their traditional competitive economic advantages; and the scarcity of investment capital (especially for Africa). After overcoming a recession at the opening of the 1990s, the industrialized nations will forge ahead ever faster and stronger, revitalized by economic and political unity and an accelerated, information-based focus on trading strength rather than military might. Increasingly, the world will be one in which developing countries must bridge a widening gap to achieve industrial status.

Yet, despite these challenges, some nonindustrial states will advance to a higher plane of economic development. There will be success stories in the 1990s and beyond—countries that will jump ahead into the rungs of industrialized nations and assume membership in exclusive clubs such as the Organization for Economic Cooperation and Development (OECD). These are countries where investment on the ground floor will prove to be immensely profitable. What are the factors, then, that will be critical for success in the next decade? What attributes will propel some countries forward, while their absence will doom others to limitless and hopeless underdevelopment?

These questions are critical to the development process as we move into the twenty-first century. While the title of this volume, *New Tigers and Old Elephants,* may appear lighthearted at first glance, the disparity between economic breakthrough and stagnation represents a crucial choice for the countries involved. Scholars may debate the precise definitions of the anthropological terms, but there is general accord that *tigers* are the countries that have risen to a more advanced economic level and that *elephants* are the upper strata of the remainder, that have continued to lumber along at the same, or even a lower, level of development but maintain a high level of competitiveness in certain sectors. Tigers are achieving their potential; elephants are not.

The goal of this book is to examine these factors and attributes to identify countries that will be winners in the development game as we enter the twenty-first century. "Winning" is when economic growth becomes largely self-sustaining, the economic base shifts from an agricultural or primary commodity base to industry and services, and socioeconomic conditions improve beyond simply meeting basic human needs. During the 1970s, winning countries were the Asian tigers: Hong Kong, Singapore, South Korea, and Taiwan. During the 1980s, this group broadened to include Chile, Malaysia, Thailand, and possibly Mexico (despite the problems in December 1994 and early 1995).

On the other side of the ledger will be elephants that just keep lumbering along; they are the countries of the future and, unfortunately, perhaps they always will be. Their collective track record has been one of high expectations, occasional but unsustainable spurts of growth, and ultimate disappointment over the lack of social and economic gains. At the same time, the mammoth size of these economies has allowed them to lumber on due to one or two sectors that are outstanding. The development path is unsteady, yet hinting at glory to be. Within this group are Brazil, India, Nigeria, and Venezuela. It is also important to clarify that not all nations are tigers or elephants; there are outright failures—such as Liberia and Afghanistan—where even the existence of a government is questionable, let alone any type of economic development.

The core of this book is divided into eleven chapters. Chapter 1 discusses the patterns of development in the 1990s and beyond and chapter 2 provides a brief discussion about the theoretical parameters of development. The next four chapters (chapters 3 through 6) defend our picks for the upcoming round of nations to emerge as tigers. They are followed by four chapters (chapters 7 through 10) on some key nations that should continue to lumber on as elephants. Finally, chapter 11 discusses how decision makers in both the public and private arenas may wish to position themselves into the opening decade of the next century.

Notes

1. *The Wall Street Journal* (14 May 1992).
2. "Third World 'farce,' neglect from industrial nations," *The Financial Times* (1 June 1992).
3. Ian Davidson, "New World Disorder," *The Financial Times* (16 March 1992).

Part I

Introduction

1

Transforming Elephants into Tigers: Development Patterns in the 1990s and Beyond

The fault, dear Brutus, is not in our stars,
but in ourselves, that we are underlings.

—William Shakespeare,
Julius Caesar, Act I, Scene 2

The path to development in the postwar era was more advantageous for some countries than others. Blessed with the right mix of resources, economic policies, and substantial support from the Western developed world, which did not wish to see communism succeed, a number of states—most notably the Asian tigers of Hong Kong, Singapore, South Korea, Taiwan—managed the takeoff to the stage of newly industrialized economy in a relatively short period of time.

The process was not easy even for these success stories. South Korea was completely devastated by the Korean War (1950–53) and forced to rebuild from the ground up while maintaining a substantial military budget to deter an aggressive North Korea. Taiwan was the last bastion of the losing side of China's civil war, and it too was forced to finance a sizeable military machine to safeguard its survival. The city-state of Hong Kong, a British colony, was threatened as an island of capitalism by the massive presence of China, the home of Maoist communism. Singapore, another city-state and former British colony, began its independent existence amidst a sea of ethnic tensions and communist insurgencies in Southeast Asia. However, the combination of cold war politics and these states' pragmatic approaches created opportunities for successful economic development that is not likely to be easily replicated in the post-cold war era.

Unfortunately for those left behind these Asian tigers, the world has changed. We believe that the next decade looms as an era of unprecedented challenges for the developing countries of Africa, Asia, the Middle East, and Latin America, which will face a substantial set of roadblocks in their drive to prosperity. The ability to transform challenges into opportunities will be the most highly prized commodity in the developing world over the next ten years. Almost in spite of themselves, some countries will indeed surmount those problems to become new tigers. This chapter will enumerate the challenges arising from our assumptions about the 1990s and beyond, and then describe the qualities needed by countries to succeed in converting these challenges into opportunities. We will construct a framework for analyzing individual countries, which will enable us to distinguish tigers from elephants in the second part of the book.

We make four critical assumptions in defining the era of the 1990s and beyond. First, after a period of slow growth in the early 1990s, the rest of the decade (and the early part of the next century as well) will be relatively good for the industrialized nations who should see real gross domestic product (GDP) growth rates that are above recessionary levels, but not more than 4 percent.[1] Powered by new technologies, a renewed focus on prosperity rather than cold war conflict, and by supranational unity, the developed world will enjoy a new wave of economic prosperity beginning in the mid-1990s and continuing into the 2000s.[2]

Second, we assume that the next decade will be a time of scarce investment capital for developing countries. The industrialized world will have heavy capital requirements internally, while competition among developing countries for external financing will sharpen. The International Monetary Fund (IMF) warns, "It is important to ensure that economic policies in the industrial world will allow financial markets to direct resources toward the reforming countries in other regions."[3] Based on our assessment of OECD capital requirements and competition for capital in the developing world, we are not optimistic in this regard. Those developing countries that desire access to capital will have to make difficult economic adjustments to be regarded as investment grade credits by international rating agencies. Without the proper ratings, many developing countries will face higher costs of raising capital; some, like most of Africa, are likely to face the prospect of virtually no capital, except of the official development assistance type.

Third, we assume that many of the traditional competitive advantages of developing nations will be lessened or irrelevant in the next ten years. Cheap labor, raw materials, and proximity to geopolitical hot spots will be of declining importance in a world devoid of cold war tensions and powered by technology rather than muscle power. This will diminish a number of traditional paths to development, making the task that much more challenging.[4]

Finally, we assume that the next decade will witness an upsurge in ethnic and racial discord, much of which will be centered in developing nations. Indeed, as journalist Robert Kaplan wrote in early 1994 of events in West Africa and how they related to the world: "West Africa is becoming *the* symbol of worldwide demographic, environmental, and societal stress, in which criminal anarchy emerges as a real 'strategic' danger. Disease, overpopulation, unprovoked crime, scarcity of resources, refugee migrations, the increasing erosion of nation-states and international borders, and the empowerment of private armies, security firms, and international drug cartels are now most tellingly demonstrated through a West African Prism."[5]

The implosion of the nation-state in Afghanistan, Liberia, and Somalia—not to mention Georgia and the former Yugoslavia—represent these trends. Internal disunity will pose a severe challenge to countries struggling to reform their economies. The extent to which nations can overcome these centrifugal forces will play a key factor in determining their success in joining the ranks of the industrialized world.[6]

The bulk of this chapter will be devoted to examining the challenges that these key assumptions pose for developing nations in the next ten years. Once these challenges have been enumerated, we can define the qualities that may enable countries to overcome them. Then and only then can we begin to outline a model for successful development in the next decade. We will begin with our first assumption: The upcoming decade will be a good one for the industrialized world.

What About the New World Order?

Paradoxically, the next "turn of the century" looks to be a good time for the industrialized nations. As Marvin Cetron and Owen Davies suggest in *Crystal Globe,* a new, more flexible world order is forming that will set the stage for a more peaceful and prosperous world for industrial countries over the next ten years. With leaders focusing on measures to promote international trade and the well-being of trading nations, rather than to suit the needs of ideological and military competition, a major military conflict will be virtually unthinkable (or so we hope!). Moreover, world leadership will be determined not by military power, but by trading strength. The postwar military powers will be replaced by three powerful and somewhat unified regional economic blocs: the European Union (EU), the Pacific Rim (in which China will slowly gain ascendancy over Japan), and North America. As the focus shifts from national security to international trading concerns, the latter part of the 1990s will be an era of growth and relative prosperity for much of the industrialized world.[7]

While crystal ballers differ in their degree of optimism, there is consensus on several fronts. In *Flashpoints,* Robin Wright and Doyle McManus highlight three major changes in the future of world power relationships. First, they concur with Cetron and Davies that economic power will be more critical than military power. Second, they suggest that power will be more widely diffused than ever, as the two-superpower model is superseded by at least four (maybe five) great powers. Also, the diffusion of power will spread beyond national governments to thousands of other institutions including regional trading blocs and multinational corporations. Third, the diffusion of influence will create a new kind of instability through the diffusion of threats.[8]

Wright and McManus agree on the huge role of technology in future power relationships. According to Enrique Iglesias, president of the Inter-American Development Bank, a "new technological paradigm holds, based on information, microelectronics and knowledge in general, while the world economy for the first time is not so narrowly dependent on labor and natural resources."[9] Technology will bring unheard of comfort and ease to many. However, this very growth and prosperity for the industrialized countries will pose challenges for those struggling to catch up. Sadly, Cetron and Davies take a much dimmer view of the prospects for developing nations. For much of the developing world, the economic and political advances of the 1990s will "go largely unnoticed."[10] They will face a new and daunting set of challenges related in large part to the good fortunes of the industrialized community. Sir Peter Leslie's claim that the developing countries will face a "period of neglect" from the industrialized north reflects the north's preoccupation with its own evolution into the post-cold war era. Countries are "voluntarily surrendering some of their hard-won sovereignty to compete for wealth in the new global economy."[11] A focus on north-north trade and integration, as exemplified by the enlarged European Union and the North American Free Trade Agreement, plus expanded ties with former Eastern bloc nations, threatens to lock developing nations out of the more lucrative global trading agreements.

There is little prospect that most developing countries can unify similarly to protect themselves against exclusion although there are efforts being made to that end, such as North Africa's United Arab Magrib and the English-speaking Caribbean's Caricom. The developing world has already fragmented into distinct groupings with different needs. The first and most needy comprises those desperately poor countries that still rely heavily on what Alvin Toffler dubs "First Wave peasant labor." The second group includes countries such as Brazil, India, and China, which are already important industrial powers in some way, but are also saddled with vast populations still scrambling for subsistence in a pre-industrial era, agricultural setting. Finally, there are the fortunate economies like Singapore, Taiwan, and South Korea that are completing the industrialization process and advancing swiftly into Third Wave high technology status.[12]

Indeed, the evolution of the new global economy appears to be widening the gap not just between rich and poor nations, but among developing countries themselves. The Latin American winners, Mexico (with a number of difficult political and economic hurdles yet to overcome) and Chile, are leaving the rest of the region behind, while the Asian tigers are in a league by themselves. The World Bank projects that despite its optimism on the long-term effects of policy reform in the developing world, the distribution of the benefits from liberalization will be uneven. Specifically, over 50 percent of these benefits will be absorbed by East Asia![13] The risk is that widening divisions and bitter disappointments may lead to a profound disenchantment with democracy and capitalism in the developing world, provoking a return to authoritarianism and state-controlled economies.

The challenges posed by these trends for developing countries cannot be overstated. Wright and McManus warn that just as the world is rushing forward toward a new golden age of global unity and prosperity, it is simultaneously rushing backward toward a "new dark era of fragmentation." New technological prowess will create new power equations based on economic strength. New economic competition will create winners and losers. In this process, ominously, "the 4 billion people in what used to be called the Third World were the most vulnerable players of all.... Entire continents are in danger of dropping out of a competition that hinges less and less on natural resources and cheap labor—some developing countries' only selling point—and more and more on high technology."[14] The danger is that the Western countries will simply lose interest in the "South," or southern hemisphere, leaving the weakest countries farther behind and widening the sociopolitical divides between the prosperous, industrialized North and the struggling South. The United Nations has estimated that by the year 2000, countries accounting for 60 percent of the world's population (around three billion people) will produce an annual per-capita average income of about $840, while Americans will produce almost $18,000.[15] As the United Nations warn: "The world can never be at peace unless people have security in their daily lives. Future conflicts may often be within nations rather than between them—with their origins buried deep in growing socio-economic deprivation and disparities. The search for security in such a mileau lies in development, not in arms."[16]

Challenges of the 1990s I: Technology

What factors will enable some countries to graduate into Second or even Third Wave levels, while others remain mired in the First Wave? The answer to this question lies largely in how well equipped each country is to meet the challenges of the 1990s. As the industrialized world forges ahead into new heights of technological competence and sophistication, the barriers to new entrants soar ever higher. According to Toffler, "[F]rom now on the world will be split between the fast and the slow."[17] He postulates that as the speed with which data, information, and knowledge race through the world economy picks up, the acceleration process leaves tortoises behind. Toffler's ideas are paralleled by Iglesias: "Future growth will be ever more dependent on the incorporation of technological know-how and innovation into the production process."[18] This means that many of the world's poorest countries could be increasingly isolated from the world economy and left to stagnate.

Challenges II: Capital Shortages

Another challenge arises from our assumption that global competition for external financing will intensify during the 1990s. A shortage of international capital for development was apparent by the late 1980s as slumping economies and inward-turning moods throughout the industrial world caused overseas investment capital to dry up. By the early 1990s, the United States was trying to regenerate its own economic growth rate, while Germany was focusing on absorbing the former East Germany. The world's richest country, Japan, was deeply shaken by its stock and property market plunge, and was embarking on a nearly $100 billion domestic spending program to reflate its sagging economy. While these are all short-term phenomena, we believe that capital shortages for developing countries will not abate.

In fact, demands on world savings will multiply from large budget deficits in several large industrial countries, the expensive reform process in Eastern Europe, and global efforts to clean up the environment.[19] Much of the limited aid and private capital that will be available will go to the north's own poor: the republics of the former Soviet Union and Eastern bloc countries. At the same time, past excesses plus increased regulatory burdens will keep commercial bank lenders, especially in the United States, cautious about lending to developing nations. Moreover, savings trends in industrialized countries are not encouraging. The trend in depleting global savings could be furthered by the gradual, yet growing, consumer demand of the populations in the East Asian tigers.

As a result, it is far from certain that external financing will be available to provide investment in developing countries. Prospects for an improvement in the availability of financing from market sources for the vast number of noninvestment grade investors are poor, and scope for expansion in official lending is limited. Thus, the combination of a relatively low savings level in industrialized countries, plus soaring investment requirements internationally, suggests that developing countries will need to fight for every dollar of foreign investment.[20] Moreover, what capital is available will be costly; the World Bank projects that a continuing shortage of worldwide capital for developing countries will produce high real interest rates, averaging over 3 percent per annum, throughout the 1990s.[21]

At the same time that worldwide demands on capital mean that nations will have to present very attractive investment profiles, we further assume that many of these same nations are losing their most important sources of strength: geopolitical location, cheap labor, and natural resources. Those exact trends that will help move the industrialized world into an era of peace and prosperity will act in some ways to retard development in parts of Asia, Africa, the Middle East, and Latin America.

Challenges III: Peace Breaks Out

One of the strangest paradoxes is the end of the cold war. In the postwar era, the cold war was often a major advantage to developing countries. Strategic and geopolitical importance to either the United States or Soviet Union was a key factor in some countries' ability to develop. Mexico, of course, jumps out at us immediately in this regard. The U.S. government viewed Mexico's descent into the debt crisis in the early 1980s as, at least in part, an invitation to left-wing forces on its southern frontier. This fear, plus the flood of unwanted Mexican immigrants across the border, prompted the U.S. government to intervene steadily and strongly on Mexico's behalf throughout the debt renegotiation process (so strongly, in fact, that at one point U.S. congressmen objected vehemently to what they viewed as subsidies granted by the U.S. to Mexico in the issuance of U.S. Treasury bonds). Even before Mexico, earlier debt crises involving Turkey and Poland in the late 1970s illustrated the benevolent role of an interested Western government in resolving financial difficulties.

A case study by Harvard Business School (*Managing the Debt Crises of Developing Countries,* 1981) highlights the importance of this practice.[22] According to the case, Turkey's serious debt problems of the late 1970s stemmed directly from the government's ill-conceived scheme to attract foreign currency deposits by promising high interest rates and currency risk protections. Not surprisingly, the scheme collapsed of its own weight, leaving the government "coffers bare." Official creditors, at first recalcitrant, rethought their position quickly as the Shah fled neighboring Iran and U.S. monitoring bases in Turkey become more critical. In the end, a deal brokered by governments of the U.S., France,

Germany, and the United Kingdom provided new capital and a comprehensive rescheduling plan for Turkey.[23]

More subtly but no less dramatically, other countries around the world have benefited out of all proportion from special Western aid and attention due to geopolitical factors. This attention arguably distorted their economies and made them overly dependent and vulnerable while obscuring the need for real reform, as in Israel, but the financial effect is indisputable. On the other side of the ledger, of course, are ex-Soviet clients such as Cuba, whose services are no longer needed in a post-cold war world.

Thus, the flip side of a world devoted to trading prosperity rather than military might is, in most cases, the end of the strategic location advantage. As the communist threat recedes, Western countries are focusing more on external economic and political integration (North American Free Trade, European Monetary Union) as well as internal challenges (the U.S. government deficit, German reunification). The end of the cold war is bad news indeed to countries like the Philippines and Cuba, which have long traded on their access to strategic locations. It should also be noted that the changed circumstances have augmented or maintained the strategic value of a small group of other nations, especially the Gulf states and Saudi Arabia. As economic concerns are driving global relations, control over Middle Eastern oil remains important—with or without the cold war.

The end of the cold war also leaves industrialized countries free to concentrate on inward-looking policies, which is even worse news for the south. One by-product is protectionism toward countries outside of the new trading blocs. Success stories of the past two decades, when the Asian tigers (South Korea, Taiwan, Singapore, Hong Kong) created miracles by exporting their way to economic prosperity, will not be repeated. New barriers to imports of textiles, footwear, and leather goods—exports that typically jumpstart developing economies—ensure this. The Bank for International Settlements (BIS), for instance, directs sharp criticism at the trade policies of wealthy nations, pointing out that a new generation of trade barriers has appeared. Moreover, the BIS charges that traditional restrictions on goods from the developing world such as farm products and textiles remain in place, while fifty-one developing countries have liberalized trade policies in the past five years and made painful changes.[24]

Challenges IV: Cheap Labor, Anyone?

New barriers to export-led growth also reflect the downgrading of cheap labor as a source of economic power. Broadly speaking, developing countries will see their comparative advantage in cheap labor erode in the coming decade. Cheap labor was the engine of growth to many developing countries in the 1970s and 1980s, helping propel the Asian tigers forward and prompting investment in developing countries from Mexico to Egypt. However, Toffler suggests that under the new, knowledge-based system of wealth creation, "cheap" labor will become increasingly "expensive." Better technology, faster and better information flows, and streamlined inventory and organization will all produce savings far beyond that which can be squeezed out of hourly workers, according to Toffler. He cites Umberto Colombo, Chairman of the European Union Committee on Science and Technology. Cheap labor, says Colombo, "is no longer enough to ensure market advantage to developing countries."[25] It will be more profitable to run an advanced facility in Japan or the U.S. with a handful of highly educated, highly paid workers than a backward factory in China or Mexico dependent on masses of badly educated, badly paid workers.

To its dismay, Mexico is already finding this to be true. Rhetoric and publicity notwithstanding, multinationals are not rushing to relocate operations from Southeast Asia to Mexico despite the much-vaunted advantages of cheap labor and free trade with the U.S. One problem is that Mexican labor may not be all that cheap. It is conceivable that the hidden costs of operating in Mexico, including high employee turnover, inefficient mail and telephone systems, and lower productivity, may overwhelm the direct labor cost savings. Indeed, the actual labor costs in Mexico when all of this is taken into account can be as high as $15 an hour, rather than the $1.50 often cited. In other words, being a low-cost supplier requires much more than just "cheap labor."

Moreover, research shows that multinationals seek more than just cheap labor in choosing a production site. Thus, most companies indicate that they will remain in Southeast Asia because of its market potential, rather than relocating to Mexico even after a free trade agreement with the U.S. is concluded. With domestic growth rates in Southeast Asia expected to accelerate during the 1990s, companies have little appetite for a move to Latin America. According to the International Finance Corporation, the World Bank's private lending arm, foreign investors no longer seek cheap labor in developing countries. Rather, they want big markets, quality infrastructure, and high skill levels. The strongest magnets for foreign direct investment are also countries with high savings rates, such as Chile and Thailand. Low labor costs have "been of diminishing importance" according to the study, largely because labor accounts for a dwindling share of overall manufacturing costs, averaging just 10–15 percent. Rather, "foreign investors are strongly influenced by structural factors such as market size, the quality of infrastructure, the level of industrialization, and the size of the existing stock of foreign investment."[26]

Challenges V: Raw Materials, Anyone?

Another major blow to developing countries is our assumption that exports of bulk raw materials will become less and less an engine of growth in the 1990s. According to the European Union's Colombo, "In today's advanced and affluent societies, each successive increment in per capita income is linked to an ever-smaller rise in quantities of raw material and energy used." Advancing knowledge permits us to do more with less, and enhances our ability to create substitutes for imported resources. This shifts power away from the raw material exporters.[27] An example cited by Wright and McManus is the decline of copper for Zambia and Chile. As communications wizards develop fiber optics rather than electrical and telephone wiring based on copper, much of the copper market is disappearing.[28] These trends are already reflected by market forces, as prices for many commodities (tea, coffee, cocoa) seem to be permanently stuck at low levels. With most developing countries heavily dependent on exports of bulk raw materials (e.g. copper, bauxite, gold), their ability to fund future development through these sales will progressively decline.

Challenges VI: Free to Fight Each Other

In addition to these external challenges, our final assumption is that countries around the world (both developed and developing) will face a new and improved set of internal challenges as well. The 1990s will be a decade of ethnic conflict at levels unheard of since World War II, released from Pandora's Box by the sudden end of authoritarian rule in much of the Eurasian land mass. The growing power of world religions, especially Islam and Catholicism, will open the door to a new round of conflicts, possibly holy wars. These divisions are infinitely worsened by the painful economic austerity programs implemented by developing world governments, with short-term drops in living standards usually the first tangible result. Some of this has already begun in the Balkans, Africa (Sudan), and the Transcaucasus (Armenia, Azerbaijan, and Georgia). Even Spain, Belgium, and Canada are infected by these forces and could be threatened as well. As Wright and McManus observe: "The challenge was being felt not only in crumbling multiethnic and multinational states like Yugoslavia, Canada, the Soviet Union and Peru, and divisive, multireligious countries such as India, Nigeria and Northern Ireland. In small ways and large, regionalism, racism, and religious rivalries were spreading or deepening even in nation-states considered homogeneous." The worldwide outburst of ethnic and rationalist movements will give way to, at best, a "messy sorting-out process" over the next decade. Some of the worst challenges may be in Latin America, where Indians, mestizos, and white descendants of European settlers have inhabited the same countries for four centuries, yet still live as different nations.[29]

At the same time that these centrifugal forces are gathering momentum, much of the developed world is working to pool resources and present a more united front. Industrial nations are surrendering power to international bodies to an unprecedented degree in the hopes of promoting the collective good. Supranational institutions like the United Nations and EC are taking on a new authority and importance. Regional trading blocs in North America, Europe, and Southeast Asia will unify and fortify the economies of the member nations. The bottom line will be an increasing differentiation between those countries that are invigorated by the unification process, and those countries that are debilitated by divisive internal conflicts. As the industrialized world consolidates its positions, once again, barriers to new entrants will go up.

Making the Grade

But, despite these challenges, countries around the world are embarked on a fast and furious drive to join the industrialized club of nations, and some will succeed. Western leaders unanimously laud the economic changes being implemented throughout the developing world, not to mention the wave of democracy sweeping through Latin America and proceeding on a more uncertain and tortuous path through Africa and Asia. A few lucky countries will graduate to new tiger status during the next decade, thanks to their possession of several key attributes. Of course, the definition of a new tiger is difficult to pinpoint. Overall growth rates are one indicator often used. The 1980s saw East Asia—including the four tigers of South Korea, Taiwan, Singapore, and Hong Kong—grow at an annual average rate of 6–7 percent, while per capita income in black Africa declined at a 2.2 percent annual rate, and in Latin America fell by around 0.6 percent per annum. Business returns on investment would be a more useful indicator, but these will vary widely by industry sector. Other descriptive characteristics used include per capita income, manufacturing as a percentage of total GDP, and manufactured goods as a share of total exports.

In this new world (dis)order posing challenges of both political and economic nature, what will make the difference between those countries that break through in the development game, and those left behind? Key factors from our view include technology power, an adequate regulatory legal system, sequencing, the right resources, and a responsible leadership elite.

Technology Power

First and foremost will be a country's ability to comprehend and use technology. Wealth creation in the 1990s will be knowledge based to an unprecedented degree. The future path to economic development and power will no longer rest as heavily on exploitation of raw materials or human muscle, as on the strength of the human mind. In this context, Toffler noted, "the most acute shortage facing LDCs is that of economically relevant knowledge."[30] This point is also underscored by the Clinton administration's Secretary of Labor Robert Reich. "Each nation's primary assets will be its citizens' skills and insights. Each nation's primary political task will be to cope with the centrifugal forces of the global economy which tear at the ties—binding citizens together—bestowing ever greater wealth on the most skilled and insightful, while consigning the less skilled to a declining standard of living."[31]

Much work, including the insightful studies of Michael Porter in his widely read *The Competitive Advantage of Nations*, has focused on the role of technology among critical success factors for nations. Porter, among others, highlights the role of successful macroeconomic management, as well as deliberate government policies to promote certain sectors. In turn, these factors are closely tied to policies that promote innovation throughout the economy, confirming once again that the ability to incorporate technological innovations is vital to a country's competitiveness.[32] Another viewpoint is presented by Sanjaya Lall, drawing on the seminal work of Albert Hirschman on economic development and the role of technology. Lall cites the OECD-defined framework for assessing national technological capabilities (NTC) in developing countries, which turns on the interplay of capabilities, incentives, and institutions. Most important, Lall suggests that human capital includes not just the skills generated by formal education and training, but also those created by on-the-job training and experience—as well as a legacy of inherited skills, attitudes, and abilities that can either aid or inhibit economic development. Literacy and primary education are essential for all forms of efficient industrialization, but as more sophisticated technologies are created, nations will need increasingly advanced, specialized skills in their work force and managerial pools.[33]

Thus, countries that take account of the new role of knowledge and the need to adequately train their work force in their development strategies will be well-positioned for the 1990s. Criteria for success here can include factors such as educational level and quality, infrastructure, standard of living, and skills of the work force. In particular, education will be essential to competing in the 1990s, propelling Argentina, for example, while acting as a drag on the Sudan. This will feed into countries' ability to mobilize their large and fast-growing populations, converting them from displaced rural laborers to productive and efficient urban dwellers.

This category goes much, much deeper than statistics on higher education and research and development spending. Entrepreneurial culture and institutional memory are enormous pluses, strengthening Lebanon's prospects but depressing Russia's and Ukraine's. Openness to external knowledge is another critical factor. Brazil, for instance, lost much ground by barring most outside computers and software during the past two decades. The country did manage to build its own computer industry, but its products were decidedly backward. This means that Brazil's banks, industries, and other businesses are using inferior technology. Brazil is coming around on this issue, but it remains a thorny one for China. The Chinese cannot cut themselves off from participation in the global economy and the telecommunications and computer networks that support it. But, how to develop a modern economy using Western ideas on technology, while excluding at the same time Western ideas on politics and democracy?

Sequencing: Money Before Politics

Despite the seeming lunacy of this proposition, the Chinese may have a point. Experience in the 1980s does suggest that countries are better able to implement meaningful economic reform under a closed political system than under a democracy. Chile, of course, is the best example of a country that underwent substantial and successful economic liberalization under a repressive political system. Indeed, there is much evidence to suggest that Chile's economic reforms would not have been anywhere near as easy or successful in a democracy, where the head of state would have had to contend with demands from opposition parties, labor leaders, and dispossessed workers. Mexico, too, managed to accomplish its economic reforms under a relatively closed political system. U.S. praise notwithstanding, Mexico is hardly a model of participative democracy.

On the other side of the ledger, countries that reformed their political system first have been largely unsuccessful so far in implementing economic reform. Virtually every country in Central and Eastern Europe, and most dramatically the former Soviet Union, illustrates this point quite effectively. The young and fragile democracies of Latin America, such as Argentina and Brazil, have also encountered enormous political obstacles to economic reform.

Viewed from this angle, the Chinese determination to implement economic liberalization without political reform may be more pragmatic than pigheaded. (The downside, of course, could be equally dramatic—another revolution or civil war.) In other countries, the very democratic trends that Western leaders are applauding may yet undermine economic reform. For many citizens of Latin America, for instance, economic liberalization has brought nothing but seemingly ever-dropping standards of living. While the new policies have generated new wealth, trickle-down effects are not significant because of the imperfections of the Latin American markets. Specifically, the region's antiquated political system and judiciary are not being reformed as fast as the economy, letting corruption go largely unpunished. As the gap between the haves and the have nots widens, the danger of the backtracking and popular discontent grows.

These examples highlight the critical role of sequencing in determining success or failure for a developing country. It appears that countries that can maintain relatively closed political systems throughout the process of economic liberalization will be better positioned than those countries with full democratic systems. In particular, countries with new and untested political freedoms will be particularly hard pressed to implement serious structural reform. Clearly, democratic governments have a more difficult path than repressive dictatorships.

One is cautioned here, however, in making the judgment that dictators make better debtors than democrats. The dismal track record of Mobuto's Zaire is one of the most blatant examples of a dictator being a really bad debtor. The key issue is a responsible political elite that can provide political stability, improve the standard of living for the majority of the population, and is not overwhelmingly venal to the point that official corruption is itself an obstacle to growth. East Asian tigers have been governed by largely authoritarian regimes that have provided a high degree of stability (with a few turbulent interludes). The tigers, while authoritarian, are not totalitarian. As Steven Schlosstein noted, "Unlike the central command systems of the Soviet Union and China, all the Little Dragons [tigers] encourage the growth of private ownership and permit a remarkable degree of personal freedom."[34] Additionally, the political institutions in these countries have a higher degree of development and are regarded as reasonably legitimate by the people—factors central to conducting economic reforms. While there have been democratic governments that have failed to implement economic reforms for many of the reasons elaborated above, many authoritarian regimes have failed for the same reasons.

Lest We Forget: The Role of Economic Reform

The issue of political reform is closely related to that of economic policy. While external factors are always important, in the end the decisive factor is a country's own mix of domestic economic programs. The IMF is quite blunt about it: "The most important requirement for a restoration of growth is an improvement in domestic policy."[35] In this context, the IMF further points out that economic performance in developing countries is often hampered by a combination of macroeconomic instability, structural rigidities and inefficiency, and high debt-service costs that absorb a high level of tax revenues and foreign exchange earnings. These problems must be addressed if adequate growth is to emerge, and each country's will and ability to do so becomes a third critical success factor. Most importantly, a country's prospects for growth over the medium term rest on its ability to strengthen investment performance. In turn, investors require a mix of policies that foster higher domestic savings, a return of flight capital, and a more efficient allocation of resources.

The ability to attract investment is often considered only with regard to a country's external creditworthiness and attractiveness to foreign investors. However, the national savings rate is also critical. Interestingly, during the early 1990s, as heavily indebted Latin American nations such as Brazil, Argentina, and Mexico rushed to the newly hospitable international capital markets to raise funds, Chile largely stayed home. Chilean officials point out that their domestic market provides longer dated and cheaper financing than the Euromarkets—a direct result of high domestic savings rates. In general, the trend rate of per capita GDP growth in high-saving LDCs is significantly higher than other countries.

Thus, although LDCs will continue to rely on external capital, this must be accompanied by measures to reduce impediments to all investments including macroeconomic instability and structural distortions. Most studies indicate that the main prerequisite in LDCs for increases in investment levels is the establishment of a stable macroeconomic environment. High inflation rates in particular have a significant negative impact on private investment. In a situation of high inflation, volatile real interest rates and exchange rates, and a large fiscal deficit, investors are likely to postpone investment plans.[36]

While "economic reform" is clearly in the eyes of the beholder, some common threads have become clear. At a conference in 1987, John Williamson listed several areas in which structural reforms have been concentrated. This list, which has become known as the "Washington Consensus," sets out the common denominators of economic reform, including:

- fiscal discipline, of which tax reform and a reordering of public spending are key priorities;
- deregulation of the financial systems and interest rates
- competitive, reasonably stable, market-driven exchange rates;
- trade liberalization;
- openness to direct foreign investment;
- privatization of state enterprises; and
- deregulation and state form.[37]

Stable macroeconomic policies are not the only requirement for success in regard to economic reform. Sequencing issues, relating to how fast, and in what order, markets should be reformed, are also critical. It has been strongly argued that impediments to capital movement should not be relaxed before the domestic financial sector is liberalized. If capital controls are eased when domestic interest rates are fixed at artificially low levels, capital flight will result (as Argentina learned in the early 1980s). In inflationary countries, an additional problem arises: Liberalization of domestic financial markets can only be fully undertaken if the fiscal deficit is under tight control. If the government deficit is not controlled when domestic interest rates are liberalized, inflation will have to increase in order for the government to collect the same level of resources. Thus, capital controls should only be lifted after the domestic financial market is reformed and interest rates raised. In turn, interest rates should only be liberalized after the fiscal deficit is under control.[38]

Moreover, even implementing the proper reforms, in the proper order, is not enough to ensure an attractive investment climate. A key determinant of the survivability of economic reform is the extent to which the reform program is credible in the eyes of outside observers. If the program is not credible and the public expects it to be reversed, then companies and consumers will actually take steps to undermine the effectiveness of the government's program. Labor markets, for instance, are generally overlooked in reform efforts, but postponing changes in this area can be costly to the overall policy, especially in regard to education related to new technology. Higher unemployment generates increased political resistance, which in turn undermines the credibility of the reforms. This suggests that labor markets should be reformed at an early stage of the structural reform effort, in order to improve their flexibility and ability to withstand massive structural shifts.[39]

Thus, the third critical success factor, ability to implement substantial economic liberalization, in turn rests on a number of key preconditions. Sequencing of the reforms is important, as is the speed with which reform is implemented. The reform program must be internally consistent and externally credible. This will be crucial in determining the degree to which the economy will or will not be reformed over the next decade. However, IMF-style standard prescription economic reform may not be the best medicine for every country, and patterns of development will differ depending on other circumstances as well.

Adequate Regulatory and Legal Systems

Directly related to proper economic policies is the need for adequate legal-regulatory systems. This means that both local and foreign investors, companies, and labor play by the same set of rules. Most fundamentally, the rules uphold the notion that a contract is a legal and binding document. In an increasingly competitive world in terms of searching for access to capital, the lack of adequate rules and regulations pertaining to commerce and finance can represent a critical factor in retarding economic development. Moreover, a working and well-understood set of laws can help streamline the process of converting an economy from inward-oriented statism to a more open and private sector-led economy. Such laws encompass repatriation of profits, bankruptcy, and foreign ownership.

An adequate legal-regulatory system ultimately benefits all players. Domestically, it provides a legal foundation for private sector growth, while enhancing the attractiveness of a particular country in terms of investment. Proper rules and regulations also safeguard all economic actors from financial fraud; they reinforce the idea that rule of law exists and underline the legitimacy of the political system.

Workers of the World, Unite!

Another important element in the recipe for success will be a country's ability to withstand the centrifugal forces now pulling some parts of the world apart. Clearly, nations that have a high level of cultural homogeneity have an advantage. As Schlosstein observed of Japan, Korea, Singapore, and Taiwan: "In each case cultural homogeneity makes the policy formulation process smoother, its implementation easier, and its enforcement 'cleaner.' Absent are the diverse, pluralistic pulls characteristic or either American or Western European culture."[40] New tigers will be marked, for the most part, by an absence of severe ethnic or religious rivalries. On the contrary, they will be characterized by relatively strong central authority and a willingness to surrender some degree of power to supranational institutions as well (dependent on the gains to be made in trade and foreign direct investment access). This mitigates against dramatic success stories for much of the Middle East and Africa and parts of Asia, such as Sri Lanka, which will experience societal turmoil throughout the decade. The big gambles here, of course, are countries like India and South Africa: Can they put their worst conflicts behind them?

The Right Resources

Also, new tigers will possess a good natural resource mix for the 1990s. The key requirement, as discussed above, will be human capital. Raw materials such as minerals and oil will be less crucial than during the past few decades. Indeed, the vulnerability implied by dependence on such resources may be a severe negative. As one economist notes, "The productive and technological model linked to natural resources is basically determined by the industrial reality of the developed countries."[41] However, a country's ability to feed itself in an era of fast-multiplying hungry mouths will be critical. One forecaster notes that there will be less arable land and fresh water per person by the year 2000 than at any time since these vital natural resources have been systematically measured.[42] Many potential tigers (e.g. India, Brazil, China, South Africa) boast both large populations and large spaces. Their ability to maximize the use of both resources will be critical. This implies mobilizing agricultural technology to achieve food self-sufficiency; maximizing energy technology and scarce resources, in many cases, to achieve energy self-sufficiency; and endowing huge rural populations with the skills needed in an urban world. This will lead to development patterns more reminiscent of Canada and Australia than the Asian tigers.

Responsible Leadership

For all the other variables to be properly taken into consideration, the right strategies to be formulated and implemented, and for the population to regard the policy direction as legitimate and beneficial, responsible leadership is critical. That responsible leadership establishes the credibility of the economic policy to provide a better standard of living for all involved in the process of improving the nation's well-being. Part of this is measurable by statistics, such as per capita income, doctors and nurses per person, and caloric intake. Part of this is not measurable, except by opinion polls that gauge the public's sentiment about the way things are: Are you better off now than you were a year ago?

In the cases of East Asia's tigers, leadership has played a key role, especially as the nation's elders have had a vision of where they wanted to lead their countries. In a largely authoritarian context, leaders such as South Korea's Park Chung Hee and his successor, Chun Duo-Hwan, and Singapore's Lee Kuan Yew, clearly identified economic frontiers, created a necessary consensus within the elite and a large segment of the population, and acted. It is often forgotten that the economic transformations that have occurred in East Asia have been partially the result of the national leadership. The leadership in these countries was able to take advantage of cultural homogeneity, was willing to search for new forms of government/business cooperation, and to place an emphasis on outward economic expansion. They also saw the importance of public education, working infrastructures, research and development, and value-added production. All things considered, the leadership of East Asia's and, in time, Southeast Asia's tigers created a highly favorable climate for investment. Although the leadership factor should not be overstated, it is highly significant in determining whether a nation has the potential to shift from a Fourth World status of underdevelopment—elephant status—to become a new tiger. Lack of consensus within the national leadership is usually reflected by a lack of consensus in society about national objectives and has undermined more than one experiment in economic reform.

Creating a Model

In our model, then, the decade of the 1990s will present a new and unprecedented set of challenges to developing countries. The new tigers will proceed along a unique development path. But one thing is certain—for developing countries to grow dynamically, "much will depend on a productive restructuring aimed at increasing their international competitiveness through the incorporation and spread of technological change."[43] Critical success factors include responsible leadership, knowledge-based skills, a commitment to deep and lasting economic liberalization, an ability to balance political reform with economic reforms, the ability to overcome ethnic or religious rivalries, and the skills to maximize food and energy resources.

Countries that are endowed with all or several of these attributes will have a good decade. Investors are already elbowing each other out of the way in the rush to get in on the ground floor of a new tiger. Flows of private capital into Latin America tripled during 1991 to over $40 billion, eight times the 1989 level.[44] The resurgence of economic growth and prospects for free trade agreements have sparked much of the activity, as investors anxiously await the coming of the next Singapore or Taiwan. Even the Mexican chill factor in December 1994 and in early 1995 will not end this quest by investors, especially in the long term.

With a wide range of countries from which to choose in our study, we used a selective menu approach. A number of nations were considered for either the tiger or elephant category, such as Lebanon, South Africa, Uruguay, Saudi Arabia, Colombia, Tunisia, and Indonesia, all of which merit their own chapters. Each member of this group demonstrates certain of the key variables that have been outlined. Considering the number of countries already chosen, the option of keeping the book to a manageable size was a consideration. The nations selected—Argentina, Morocco, China, Slovenia, Vietnam, Kazahkstan, Brazil, Venezuela, Peru, India, and Nigeria—are meant to be representative of either tigers or elephants.

Another factor in selection was the fact that the assembled countries have all sought to implement significant economic reforms in recent years. Moveover, as a group they also have witnessed considerable political change. In some cases, change has been the birth of the country, the fall of authoritarian regimes, and the establishment of open political systems. Another consideration is the effort to create political economies that are more integrated in the international system then in previous periods.

A final factor was our decision to largely omit the marketizing economies of the former Soviet Union and Eastern Europe. Most of the Eastern European states have existed as countries for a considerable period of time and several, such as the Czech Republic, increasingly have found that they have more in common with industrialized Western Europe. Slovenia and Kazakhstan, however, are examined as some of the stronger tiger prospects among the sovereign states that were newly created (or recreated) out of the former Soviet empire.

Our effort to identify the new tigers is based on these critical success factors. We have tried to assemble a coherent model based on data in many key areas, but in the end assertions such as these rest heavily on subjective assessments. Why is Argentina a new tiger and Brazil an old elephant? This is not a scientific process, nor do we purport to be providing forecasts or even indications. Rather, our analysis provides opportunities for speculation, which may be utilized by those in a position to invest for potential heavy gains or losses. After all, the correlation of high returns with high risks is one of the basic lessons of finance. Indeed, studies show that political risk has not deterred investors when the potential returns are considered high enough. One factor that does attract potential investors, though, is the presence of a substantial existing stock of foreign capital. Those investors who manage to identify a new tiger before the herd mentality sets in are posed for substantial and long-term benefits as a result.

References

Bradford, Colin I, Jr., ed. *Strategic Options for Latin America in the 1990s.* Paris: OECD Development Centre, 1992.

Cetron, Marvin and Owen Davies. *Crystal Globe.* New York: St. Martin's Press, 1991.

Dodwell, David. "Investors turning away from cheap labor countries." *The Financial Times* (25 May 1992).

Edwards, Sebastian. "Sequencing Issues in Economic Reform."

Fidler, Stephen. "Latin America's New-Found Allure." *The Financial Times* (19 February 1992).

Goodman, Allan E. *A Brief History of the Future: The United States in a Changing World Order.* Boulder, Colo.: Westview Press, 1993.

Harvard Business School. *Managing the Debt Crises of Developing Countries.* Cambridge, Mass.: Harvard Business School, 1981.

Iglesias, Enrique. *Reflections on Economic Development: Towards a New Latin American Consensus.* Washington, D.C: Inter-American Development Bank, 1992.

International Monetary Fund. *World Economic Outlook, May 1991.* Washington, D.C.: International Monetary Fund, 1991.

______. *World Economic Outlook, May 1994.* Washington, D.C.: International Monetary Fund, 1994.

Kaplan, Robert. "The Coming Anarchy." *The Atlantic Monthly* (February 1994): 45.
Lehner, Urban C. "Belief in an Imminent Asian Century Is Gaining Sway." *The Wall Street Journal* (17 May 1993).
MacDonald, Scott, Uwe Bott, and Janes Hughes, eds. *Latin American Debt in the 1990s: Lessons from the Past and Forecasts for the Future.* New York: Praeger Publishers, 1991.
Organization for Economic Co-operation and Development. *OECD Economic Outlook, June 1994.* Paris: OECD Publications, 1994.
Porter, Michael E. *The Competitive Advantage of Nations.* New York: The Free Press, 1990.
Reich, Robert. *The Work of Nations: Preparing Ourselves for 21st Century Capitalism.* New York: Alfred A. Knopf, 1991.
"Rich Nations are Criticized." *The New York Times* (15 June 1992).
Schlosstein, Steven. *Asia's New Little Dragons: The Dynamic Emergence of Indonesia, Thailand, and Malaysia.* Chicago: Contemporary Books, Inc., 1991.
Teitel, Simon, ed. *Towards A New Development Strategy for Latin American: Pathways From Hirschman's Thought.* Washington, D.C.: Inter-American Development Bank, 1992.
Toffler, Alvin. *Powershift.* New York: Bantam Books, 1990.
United Nations. *Human Development Report 1994.* New York: Oxford University Press, 1994.
World Bank. *Global Economic Prospects and the Developing Countries.* Washington, D.C.: World Bank, 1992.
Wright, Robin and Doyle McManus. *Flashpoints: Promise and Peril in a New World.* New York: Fawcett Columbia, 1991.

Notes

1. For example, the International Monetary Fund projects the industrial countries' average output at 2.4 percent for 1994 and 2.6 for 1995. International Monetary Fund, *World Economic Outlook May 1994* (Washington, D.C.: International Monetary Fund, 1994), 12.
2. These viewpoints are expressed in two excellent works: Alvin Toffler, *Powershift* (New York: Bantam Books, 1990) and Marvin Cetron and Owen Davies, *Crystal Globe* (New York: St. Martin's Press, 1991). Both Toffler and Cetron and Davies believe that the end of superpower conflict in the post-cold war era opens up new opportunities for progress in the industrialized world. The authors suggest that free to concentrate on economic prosperity rather than the threat of communism, OECD countries will be empowered to forge ahead more quickly and efficiently in their economic development.
3. Data on world capital shortages are drawn from a variety of sources, including studies by the World Bank and the IMF. The best exposition is in the International Monetary Fund, *World Economic Outlook, May 1991* (Washington, D.C.: IMF, 1991).
4. In chapter 30 of *Powershift,* "The Fast and the Slow," Toffler discusses his belief that "many of the world's poorest countries will be isolated from the dynamic global economy and left to stagnate" (394). In his view, "the market for strategic locations in LDCs" (395) is over. In addition, he points out that the world has

moved "beyond raw materials," (398) and that "cheap labor is increasingly expensive" (p.398) in a modern, technology-based global economy.

5. Robert Kaplan, "The Coming Anarchy," *The Atlantic Monthly* (February 1994): 45.
6. A number of observers, including Toffler and Cetron and Davies, have warned that the 1990s will not favor those indulging in costly internal strife. Indeed, in chapter 25 of *Crystal Globe* ("Ceding Sovereignty for the Global Good"), Cetron and Davies discuss the growing world tendency to cede power to supranational authority in order to advance certain, frequently economic, goals. In this context, the trend toward internal disunity in countries such as Peru and Canada can only be viewed with great alarm.
7. Cetron and Davies, *Crystal Globe,* 4-5
8. Robin Wright and Doyle McManus, *Flashpoints: Promise and Peril in a New World* (New York: Fawcett Columbia, 1991), 17-18.
9. Enrique Iglesias, *Reflections on Economic Development: Towards a New Latin American Consensus* (Washington, D.C: Inter-American Development Bank, 1992), 49.
10. Cetron and Davies, *Crystal Globe,* 4-5.
11. Wright and McManus, *Flashpoints,* 26-27.
12. Toffler, *Powershift,* 391.
13. World Bank, *Global Economic Prospects and the Developing Countries* (Washington, D.C.: World Bank, 1992). The study warns that the distribution of the benefits of trade liberalization in rich countries would be uneven. East Asia would probably win more than 50 percent of the additional export orders, with Latin America taking only about 20 percent, and Eastern Europe lagging way behind with barely 7 percent of the new markets.
14. Wright and McManus, *Flashpoints,* 7.
15. For more such data, see United Nations, *Human Development Report 1994* (New York: Oxford University Press, 1994).
16. United Nations, *Human Development Report,* 1.
17. Toffler, *Powershift,* 397.
18. Iglesias, *Reflections on Economic Development,* 84.
19. Industrialized countries with large budget deficits that require ongoing financing in international markets include Canada, the U.S., Italy, Sweden, Belgium, and Finland. See *OECD Economic Outlook, June 1994* in statistical tables pertaining to general government financing balances.
20. International Monetary Fund, *World Economic Outlook, May 1991* (Washington, D.C.: Internationals Monetary Fund, May 1991).
21. World Bank, *Global Economic Prospects and the Developing Countries.* The World Bank study further finds that concessional finance will be scarce, with only the strongest economies able to tap private capital markets, while real interest rates will stay high at about 3 percent.
22. Harvard Business School, *Managing the Debt Crises of Developing Countries* (Cambridge, Mass.: Harvard Business School, 1981). The case study discusses the interaction of government and private lenders in untangling debt crises of the nineteenth and twentieth centuries. Analysis of the case suggests that when urgent strategic interests were involved, as in the Turkish debt crisis of 1977-79, the involvement of Western governments can produce an outcome that is broadly beneficial to debtor nations. Questions are raised about international responsibil-

ity for debt crisis management in countries such as Brazil, where strategic importance is less compelling.

23. Ironically, Turkey fell into serious debt management problems in 1994 due to large budget deficits and was forced to seek the assistance of the IMF and World Bank.
24. "Rich Nations are Criticized," *The New York Times* (15 June 1992).
25. Toffler, *Powershift,* 406.
26. David Dodwell, "Investors turning away from cheap labor countries," *The Financial Times* (25 May 1992). Citing a report by the International Finance Corporation of the World Bank, the article calls big markets, quality infrastructure, and high skill levels more important as magnets for foreign investors than cheap labor in developing countries.
27. Toffler, *Powershift,* 404.
28. Wright and McManus, *Flashpoints,* 39.
29. Ibid., 48.
30. Toffler, *Powershift,* 409.
31. Robert Reich, *The Work of Nations: Preparing Ourselves for 21st Century Capitalism* (New York: Alfred A. Knopf, 1991), 3.
32. Michael E. Porter, *The Competitive Advantage of Nations* (New York: The Free Press, 1990). The work of Porter and others in this area is well presented by Iglesias, *Reflections on Economic Development,* 90–95.
33. Sanjaya Lall, "The Rule of Technology in Economic Development," in Simon Teitel, ed., *Towards A New Development Strategy for Latin American: Pathways From Hirschman's Thought* (Washington D.C.: Inter-American Development Bank, 1992).
34. Steven Schlosstein, *Asia's New Little Dragons: The Dynamic Emergence of Indonesia, Thailand, and Malaysia* (Chicago: Contemporary Books, Inc., 1991), 7.
35. IMF, *World Economic Outlook, May 1991.*
36. Ibid.
37. Iglesias, *Reflections on Economic Development,* 49.
38. Sebastian Edwards, "Sequencing Issues in Economic Reform," 1–2.
39. Ibid.
40. Fernando Fajnzylber, "Technical Progress, Competitiveness, and International Change," in Colin I. Bradford, Jr., ed, *Strategic Options for Latin America in the 1990s* (Paris: OECD Development Centre, 1992), 116.
41. Allan E. Goodman, *A Brief History of the Future: The United States in a Changing World Order* (Boulder, Colo.: Westview Press, 1993), 91.
42. This comment was made in regard to Latin America, but has significance for the entire developing world. Iglesias, *Reflections on Economic Development,* 85.
43. Schlosstein, *Asia's New Little Dragons,* 17.
44. Stephen Fidler, "Latin America's New-found Allure," *The Financial Times* (19 February 1992).

2

The Issue of Development

*The charge against economists is,
above all, that of irrelevance.*

—Peter Donaldson[1]

The issue of development is a major topic of debate, having both political and economic implications. After all, the study of development examines the flow of history—the ability of nations, peoples, and transnational actors to initiate or contend with change, to better their positions or living standards, and to compete in the global economic system. These abilities were as important in the sixteenth century as they are now. Because the selection of a "development strategy" implies a political message of preference, the study of development has gone through a number of stages of often heated debate about the superiority of one path over another. In the 1990s, it appears that the proponents of capitalism (free market economics) have won the debate over their erstwhile competitors—the Marxist-Leninist school. But what is free market capitalism or for that matter, Marxist-Leninist economics? It is important to define our terms and provide a brief review of the ideas that shaped the development of the modern world to the present and will impact the future.

Modern capitalism did not appear overnight or by the wave of a magic wand to end the Middle Ages and begin the Industrial Revolution. As George Dalton noted, "Modern capitalism had its origins in the growth of commercial foreign trade, and modern economics in the writings of the mercantilists, men who wanted to convince their kings that great wealth and material advantage to the state and nation lay in encouraging such trade and controlling the money flows that accompanied it."[2] Simply stated, development for the European powers was defined by colonies and empire—colonies that existed in a closed trading system for the benefit of the mother country.

The emergence of economic liberalism, however, moved beyond mercantilism. Probably the best known of the economic liberals is Adam Smith. In 1776, the Scottish economist was the first to produce a systemic rationale that provided the foundation of a laissez-faire ideology. Simply stated Smith, in his *Wealth of Nations,* thought the market should be allowed to function without the intervention of the state because self-interest in a free society would lead to the most rapid progress and growth a nation was capable of achieving. This meant the individual should be left free to pursue and advance his own ideas. With the market functioning in its natural course, people would have the capability to pursue an improved standard of living. Smith summarized his views as follows: "Little else is requisite to carry a state to the highest opulence from the lowest barbarism, but peace, easy taxes, and a tolerable administration of justice."[3] The notion of economic choice carries with it the concept of political freedom, hence the emergence of the democratic-capitalist idea. (In this framework, the role of the state is largely noninterventionist except for upholding the laws and protecting property rights.) Man's historical evolution was to be progressive and dynamic, linked to free markets.

Smith's perception of development was based on the ideas of specialization in production, the division of labor, and the security of property. With the accumulation of capital, the natural progress of opulence advances from agriculture to manufacturing to commerce, and the affluent society exhibits prosperity in all three areas.[4] With progress in agriculture, the growth of towns is stimulated, which in turn provides a larger market for agricultural goods, and a more developed urban and rural society offers widened opportunities for trade and shipping. In turn, enlarged trade further galvanizes manufacturing and specialized agriculture for export. At the same time, the population increases as productivity is augmented, facilitating even wider market expansion due to heavier demand and stimulating greater specialization and capital accumulation. Through such a process the economy achieves ever higher levels of development. The social order obviously benefits from the process. To Smith the process of economic development centered on the creation of wealth, initiated by the individual, and on the natural maintenance of an orderly market equilibrium between supply and demand. Smith's model was forward-looking as natural liberty enjoyed by individuals always moved the equilibrium point toward opulence.

Smith's school of classical economics was expanded by a number of other economists, two of the most famous being David Ricardo and Thomas Malthus. The former was especially important due to his ideas concerning comparative advantage and his use of words and arithmetic to extract the purely market forces of cost and demand. To Ricardo, free trade was the road to economic well-being, and comparative advantage created a natural equilibrium from which everyone benefitted. This meant that as long as it cost less to produce cloth in England than it did to produce wheat, compared to the costs in other countries, it would benefit the English to shift their resources to cloth manufactures, export cloth, and import wheat from other countries. Those other countries, in turn, will find that their comparative advantage is in the production of wheat, in which they will then concentrate their resources, while importing cheaper English cloth.

While Adam Smith is often regarded as the founding father of the laissez faire ideology as it has come to be known today, the opposite current was inspired by Karl Marx, Frederich Engels, and Hegel, and further elaborated by V. I. Lenin, and even later with Mao Zedong. The Marxist-Leninist school commenced in the nineteenth century and was born of frustration with the negative social effects of the industrial revolution. In particular, as Europe shifted from agrarianism to industrialism, traditional societies were uprooted and a pronounced shift in population patterns was characterized by rapid urbanization. At the same time, scientific progress, the improvement of machinery and productivity, and the demand for more labor stimulated a number of problems. These included child labor, long work hours, unsafe work conditions, and class differences between a wealthy and relatively small capitalist group and a much larger and poorer lower class of workers.[5]

Marx found these conditions disturbing and sought to create an alternative noncapitalist view of the future. Marx, borrowing from Hegel, argued that history was dialectical—that throughout the evolution of humankind there was at some point a force and counterforce that collided. European society had been feudal with two clear-cut groups, the nobility and the serfs. Feudal society was eventually undermined by capitalism, which created the bourgeoisie. The bourgeoisie, a new middle group of merchants, traders, and professionals, however, were to be gradually undermined by their own capitalist nature. Because capitalism required a work force to fuel industrial growth, the proletariat emerged. In time, ownership of the means of production within the capitalist system would narrow into an increasingly smaller group of large capitalists. The bourgeoisie would disappear as capitalists devoured their own and the ranks of the workers expanded. Eventually, the workers of the world would unite and rise up against capitalistic exploitation, at which point humanity would achieve a classless and egalitarian, or "communist" society. This sweeping away of the old order would be accomplished by revolutionary means (possibly violent). The first revolutions, according to Marx, would happen in the more industrialized societies where the workers were the most organized and the discrepancies between capitalists and the masses were the most pronounced. To this Lenin would add that in those places where the workers were not as numerous, as in Russia, the party could function as the vanguard of the proletariat.

The Marxist-Leninist school also maintained that their doctrine was the only proper scientific theory of any society and of socialism in particular. As one long-time Soviet economics analyst noted: "Even more important is their claim on the specificity of socialism: in contradistinction to capitalism it does not come about and function spontaneously. It must be brought about by conscious action, planned and managed following the objective laws of society as discovered by Marxism-Leninism."[6] For Lenin and others who grappled with the questions of economic development in the early years of the Soviet Revolution this hard-core belief provided them with a sense that because of the scientific nature of their predictions, they would eventually be vindicated.

The Marxist-Leninist school regarded the issue of development in Africa, Asia, and the Americas as an outgrowth of global capitalism. The capitalist machine needed both raw materials and markets, which forced it to expand overseas in the form of colonies (as in Africa and Asia) and neocolonies (as in Latin America where the states were independent but economically dominated). Imperialism also carried with it rivalries which the Marxist-Leninist school observed as part of the process of concentration of wealth in the hands of a few. Imperialistic rivalries would result in conflicts between capitalists which would eventually weaken their hold over the proletariat and open the door to world revolution.

The Development Debate Evolves: Keynes, Stalin, and Beyond

While the discussion here of laissez-faire capitalism and Marxism-Leninism is relatively simple, the point is that these two schools of thought were important ideological lenses through which modern-day policymakers and leaders came to look upon the developmental process in their respective countries, ranging from the Americas to Europe to Japan as well as their colonies. In the middle ground between the two ideologies in their purest forms were a number of variations, including social democracy and anarchism (more on the extreme part of the range). Despite these challenges, capitalism and the laissez-faire idea survived the First World War. At the same time, Marxism-Leninism was given an opportunity to create a communist society with the success of the Bolsheviks in Russia in 1917. Significantly, the rival views of development became a contest not only of ideas, but also of systems based on real resources. The duel between Soviet communism and Western capitalism that commenced in earnest in 1917 would last throughout most of the century until 1 January 1992 when the Soviet Union voluntarily dissolved itself.

The ideological confrontation between the Soviet-led East and United States-led West did not become full-blown until after the Second World War. Consequently, in the interwar period the tone of the debate was initially focused on capitalist Europe versus the struggling Bolshevik regime in the Soviet Union. The Great Depression in the late 1920s and 1930s was gleefully observed by Moscow as portending the imminent collapse of Western capitalism. In the dark days of the Great Depression capitalism underwent a subtle transformation. John Maynard Keynes (1883–1946), the author of *The General Theory of Employment, Interest, and Money* added a new element to the development debate, providing the capitalist side with new policy options and laying the foundations of what is now called macroeconomics.

Keynes was deeply perplexed by the depression's length and the inability of classical economics (as started with Adam Smith) to explain its duration. According to the classical, school wage rates and the rate of interest should fall during the downturn in the trade cycle. Both eventually reached levels low enough for businessmen to observe a substantial improvement in the profitability of new investments. These investments soon generated employment and new earnings. As this process took hold, the economy resumed an expansionary course until rising prices in the boom brought the next phase in the cycle. Keynes argued that instead of cutting labor's wages as urged by classical economists in order to restart the economy, the state (in the absence of other stimuli) should assume the businessmen's function and spend money on public works. This "pump-priming" of the economy by the government would reduce unemployment and place income in the hands of labor, who would then begin to spend again. Hence, state investment in the economy was an essential factor in overcoming obstacles to stimulating growth. This also could mean that the government's budget did not have to be balanced year in and year out, a departure from classical economics.

Keynesian economics was adopted by a number of countries in the mid-1930s and continued to be a force in the postwar period, especially in the United States and the United Kingdom.[7] Many aspects of the Roosevelt administration's New Deal in the United States reflected a strong Keynesian imprint. More important in terms of this study, Keynesian economics became a guiding force in Western thinking about development, which was often the lens through which the development issue was approached in the former colonial regions in Africa, Asia, and the independent states in much of Latin America and the Caribbean. In particular, Keynes indicated that the state could play an expanded role in the economy; that developing countries had to be wary of the potential damage that could be wrought by financial systems in capitalist economies; and that without careful management the supply of money could disrupt economic growth seriously.

An offshoot of the Keynesian school was the monetarist school, which is often associated with Milton Friedman of the University of Chicago. While Friedman is known for his contributions to the theory of distribution (i.e., that high incomes are regarded as a reward for taking risks), his major work was the development of monetary theory based on the quantity theory of money. The fundamental premise is that there is a close correlation between money supply and the level of economic activity. On the basis of this relationship, the monetarists argued that tightly disciplined control of the money supply offers a better opportunity of achieving economic stabilization than the "meddling" of Keynesian techniques.[8]

The issue of development in what was to become the developing world did not emerge in a major fashion until the end of the Second World

War. The Soviet Union and the United States, which emerged as the dominant superpowers, both favored the end of the colonial empires. Moreover, the Soviet Union had moved from its earlier stance of "socialism in one country" to the assistance of world revolutionaries in Spain, China, and elsewhere in the 1930s, to active international interest with World War II and its immediate aftermath. Greater U.S. and Soviet activity in world affairs and the end of British, Dutch, French, and eventually Portuguese colonial empires suddenly brought the issue of development front and center. This was especially the case as the U.S. was the world's leading capitalist power and the Soviet Union the undisputed flag bearer of world revolution. The two offered competing approaches to the management of the economy, the structure of political life, and ultimately how each society would evolve.

The Soviet approach to development that emerged in the late interwar period and was rejoined at the end of the Second World War emphasized the state assuming the commanding heights of the economy, state planning, and almost complete state control over distribution. The Soviets, especially under Joseph Stalin, had embarked upon central planning, heavy industrialization, and collectivization of the agricultural sector.[9] The overriding concern of the Stalinist model was "the plan," which embodied the economic will of the party and government—not profit and loss considerations.[10] Politically determined priorities reduced the role of prices to a minimum and acknowledgement of consumer demand was virtually nonexistent. In this model, private property and the private sector (with very few exceptions) did not exist.

While Stalin was able to push through a process of industrialization, which was continued by his successors, the social and environmental costs were high. Dissidence was not allowed and clearly political freedom was not a concern as reflected by the gulag system and periodic purges (especially under Stalin). However, the Soviets transformed their economy from being highly agrarian into one of the world's more industrialized states and largest economies in terms of sheer size.

For much the developing world the Soviet model held a degree of attraction. It appeared to vindicate the idea that a path of development existed where the state could be harnessed to cut through distributive bottlenecks and overturn the often rapacious business practices of the local private sectors. Central planning and five-year plans were adopted in many developing countries, a legacy that still exists in parts of Africa,

Latin America, and the Middle East. Moreover, the Soviet Union quickly positioned itself as a force sympathetic to the liberation struggles of local African or Asian independence movements. Moscow became the capital for "Third World" revolutionaries who were either trained or educated in the Soviet Union to eventually return to their homelands with a certain, Soviet-inspired view of development.

The Soviet vision of development was soon challenged by another force within the Marxist-Leninist tradition—Maoism. The Chinese Revolution, which ultimately brought Mao Zedong and the Chinese Communist Party (CCP) to power in the late 1940s, derived from different circumstances that called for a Sinofication of Marxism-Leninism. In particular, the proletariat was a small group in China, then a largely rural and peasant society. Maoism opened the door to a wider revolutionary experience that tapped the potential of the countryside. It was initially felt that agricultural labor could easily be shifted to industrial construction.[11]

China's new leadership came into the post-Second World War period closer to the revolutionary fire than the Soviets. This allowed Mao to assert in 1958 that China's "poverty and blankness" makes possible the achievement of Communism long before the insidious corruptions of capitalism and revisionism set in.[12] Mao and the CCP advocated a violent new brand of world revolution and at home sought to make human will transcend the problems of economic development as mirrored by the abortive rapid industrialization effort in the Great Leap Forward (1958-1959). In the 1960s and into the 1970s, the Chinese also retained elements of the Stalinist model of central planning, which contributed to a list of growing differences with the Soviets in the aftermath of Stalin's death.

Both the Soviet and Chinese revolutionary experiences had an impact on a world awakening in the South. Moscow's and Peking's claims that rapid development was possible (and indeed desirable) were attractive to African and Asian leaders who inherited colonial systems and were now confronted with the awesome responsibility of managing an economy and conducting political life. In Latin America, where capitalism (in various forms) had held sway for a long period, these ideas also had an influence. At the same time, many dimensions of the Soviet and Chinese experiences did not fit local conditions. Furthermore, the violent excesses of these two revolutionary experiences and their authoritarian mode of politics did not appeal to every developing country leader.

The U.S. provided another pole of development thinking. The doctrine of "the stages of economic growth" was born from the emergence of the cold war in the 1950s and 1960s and the related competition for the allegiance of newly independent countries. The most vocal proponent of this school was economic historian W. W. Rostow. In *The Stages of Economic Growth* (1961), he argued that the industrialized countries had at various times in history passed the stage of "take-off into self-sustaining growth" and that all underdeveloped countries have the potential to advance to a "developed" destination, but must go through five clear-cut historical phases. These phases are the traditional society, the preconditions for take-off, the take-off, the drive to maturity, and the age of high mass consumption. Among the key elements for reaching the take-off stage are the existence of an entrepreneurial elite accompanied by radical change in agricultural techniques and market organization.[13]

For those who subscribed to Rostow's view, the problem of development was identified as raising resources on a scale sufficient to achieve a level of investment that would change the structure of the economy and set the stage for economic takeoff. It is important to clarify that Rostow also felt that government had a positive, though not dominant, role to play in the development process. It was also implied, especially in the cold war atmosphere, that Western-style democracy was an important element of this model, perhaps a result of finding "takeoff" and a higher degree of economic maturity.

While the U.S. model provided an alternative to the Soviet and Chinese models, the U.S. frequently found itself supporting local anticommunist forces, who more often than not were lacking democratic credentials. The U.S. message about the benefits of democratic capitalism was greatly complicated by cold war concerns of containing Soviet and Chinese communisms. The trade-off of liberal democratic policy objectives for strategic concerns detracted from the capitalist model's strong points that under the right conditions it worked.

The stages of economic growth paradigm was also weakened by the differences between theory and practice. The emphasis on the mobilization of savings and investment while important, assumed the existence of other conditions that enhance development strategies based on the existence of savings and investments. In particular, Rostow and other stages theorists (such as Evesey Domar of the United States and Sir Roy Harrod of the United Kingdom) assumed the existence of such things as well-integrated commodity and money markets, highly developed transport facilities, well-trained and educated manpower, an efficient government bureaucracy, and the motivation to succeed.[14] Frequently in the developing world many of these conditions were missing. Nonetheless, cold war tensions contributed to making the quest for economic take-off a core item in U.S. and Western development strategy vis-à-vis the developing world, especially as it advanced a capitalist-oriented model that provided for a dynamic role for the private sector.

The East-West rivalry stimulated the interest in a "Third Way" of development policy. By the late 1950s, people concerned with development in Asia, Africa, and Latin America found that the acceptance of either the Soviet-Chinese or U.S. models came with external political linkages. As Nigel Harris noted: "The globe had become divided between two apparently terrifying alternatives, Washington and Moscow, capitalism and what many people supposed was socialism, the first and second worlds. The Third World, newly created from the wreckage of the old, offered a different path for humanity, a third alternative."[15] That alternative included nonalignment with either the East or West blocs in the international sphere and the option to pursue a "mixed economy" that included elements of both socialism and capitalism.

Despite the existence of differing approaches to development, the dominant view in the 1950s and 1960s was that the major objective of the national economy was to generate and sustain a substantial annual increase in gross national product. Michael Todaro noted, "For example, the 1960s and '70s were dubbed the Development Decades by resolutions of the United Nations, and 'development' was conceived largely in terms of the attainment of a 6 percent annual target growth rate of GNP."[16]

The idea of development as measured by annual increases in GNP and stimulated by free trade came under attack in the 1970s. Serious questions were raised about the applicability of the capitalist development model and, in particular, why little appeared to be changing in terms of income distribution. One of the first shots across the bow of "orthodox" economics came from Oxford University economics professor Joan Robinson, who stated in 1971, "We have nothing to say on the subject [income distribution] which above all others occupies the minds of people whom economics is supposed to enlighten."[17]

Robinson and others contended that free trade theory, which purported to be objective economics of universal applicability, was nothing of the sort. It was, in fact, "the product of thinking in white, nineteenth-century, relatively advanced capitalist economies. Although the inventors of the doctrine may not have realized it, it was heavily biased in favour of the rich industrial nations. And it continues to be so today."[18] Equally significant, Donaldson, Robinson, and others found the emphasis placed on high growth rates inaccurate as a measure of development. It was argued that behind the veneer of a rapidly expanding economy, there were substantial problems with chronic unemployment and underdevelopment; tremendous and widening disparities between rich and poor; and population growth exceeding any Keynesian techniques of pumping in more demand. Considering all of these factors, it was argued that mere growth did not equal development.[19]

A more radical critique of neoclassical economics was that of dependency theory, which was particularly strong in Latin America in the 1960s and 1970s. A number of its chief proponents included Celso Furtado, André Gunder Frank, and James Petras. Deriving from a neo-Marxist tradition, the overarching analysis of development depicted a world divided into center (or metropole) and periphery. The center encompassed the industrialized nations of North America, Europe (often including the Soviet Union), and Japan. The periphery was the developing world in Asia, Africa, and Latin America, also sometimes referred to as the South. Using the Marxist approach of stating relations of exploitation in terms of social classes, dependency theorists maintained that in the South, the bourgeoisie functioned as an extension of the center, watching out for metropolitan interests, which are usually commercial and exploitative. At times this has meant alliances with the local military, who are armed by the North and have the force to stop the so-called popular sectors or masses from achieving the goals of an egalitarian society. André Gunder Frank provides an example of this line of thought: "Underdevelopment in a dependent region such as Latin America cannot be understood except as the product of a bourgeois policy formulated in response to class interest and class structure, which are in turn determined by the dependence of the Latin American satellite on the colonialist, imperialist metropolis."[20]

Dependency theorists also argued that what was needed was a strong state to command the economy and the strategic path to "de-linking" from the North was through import substitution. Import substitution was to promote local development of products that were imported from the center. By achieving local production the linkages of dependency would be disrupted. The Third World then would be able to take charge of its development without the hinderance of forces that sought to maintain it as a source of cheap labor and raw materials as well as markets for value added goods. This also implied that local bourgeoisie would be displaced by governments truly reflective of and responsive to the demands of the popular sectors.

A more revolutionary wing of the dependency school argued that revolutionary violence was justified given the need to remove the bourgeoisie from power, especially considering their avid and self-indulgent prostitution to the neocolonialism of the metropole. In fact, to Gunder Frank, the only hope was "armed revolution and the construction of socialism."[21]

Although the dependency model was not widely subscribed to in the developing world, elements of it were popular in intellectual circles. The import substitution strategy, however, was accepted throughout Latin America, Africa, and parts of Asia. While local industries were protected behind tariff barriers and nontariff barriers to produce goods that were imported, many countries discovered that they required foreign inputs and capital to launch import substitution programs. The developing countries soon found themselves building up large external debts and creating inefficient state-run enterprises that were often characterized by bloated personnel rolls and poor-quality goods. In many cases, industry remained underdeveloped and inefficient, producing high-cost, low-quality goods behind high tariff walls. Another problem with the heavy, state-oriented economic model was that comprehensive development plans often preempted the country's supply of foreign exchange and credit, depressing economic activity outside the plans.[22]

While dependency theory remained intellectual chic into the early 1980s, one of the undergirding elements of that school of thought was eroding—Third Worldism. One factor was that a small set of economies, encompassing Hong Kong, Singapore, South Korea, and Taiwan, were rapidly taking off. Following a flexible export-growth model (not an import substitution strategy), these little tigers began the fragmentation of Third Worldist ideology. As Harris commented, by the mid-1980s the Third World "no longer was seen as a political alternative and merely denotes a group of countries—referred to more pompously as the less developed countries."[23]

The Tigers Arrive

In the mid- and late 1980s attention began to focus on what were referred to as Asia's tigers—Japan, Hong Kong, South Korea, Singapore, and Taiwan. The development breakthroughs achieved by first Japan and then the smaller countries were clearly amazing. Beginning in the post-World War II period well behind much of Latin America and Africa, these parts of Asia rapidly moved into export-led growth. By the 1980s, Japan was a world economic leader and rival to the United States and Europe. Japan, Korea, Hong Kong, Singapore, and Taiwan were called "trading states." Richard Rosecrance defines these as such: "States, as Japan has shown, can do better through a strategy of economic development based in trade than they are likely to do through military intervention in the affairs of other nations."[24] The emphasis trading states place on military spending can vary considerably as reflected by such diverse cases as Iceland, which is largely limited to a coast guard, to South Korea and Taiwan that have substantial military establishments to the mid-position of Japan. The trading state school argued that much of dependency theory was wrong and that the Soviet-led side of the development debate was deficient as it failed to identify the dynamic side of capitalism in the late twentieth century. In the late 1980s and early 1990s Korea achieved balance of payments surpluses, substantially reduced the burden of its external debt, and gradually improved the wages of its workers. In fact, the average annual GNP growth rate of Korea from 1965 to 1989 expanded at a dynamic 7 percent, compared to -1.0 percent for Venezuela, 3.5 percent for Brazil, and 1.8 percent for India.[25] In terms of real GDP growth, Korea's performance was equally impressive—in the 1965-1980 period it was an average of 9.9 percent and in the 1980-1989 period it was 9.7 percent.[26]

Korea's dynamic development from 1960 to 1989 overtook that of most Latin American, Asian, and African nations. However, its development path of export-led growth and its political economy of quasi authoritarianism-capitalism were initially criticized and held up as capitalist facades propped up by Western foreign aid. An example of this was Joan Robinson in 1979: "In spite of formidable competitive power in exports, South Korea has never succeeded in achieving balanced trade; the economy is maintained by loans and subventions and the government has to rely on its 'strategic value' for continued support from the USA."[27]

The little tigers emerged for a number of historical reasons and, to some extent, because they adopted an export-growth model that provided a high degree of innovation (à la Schumpeter). A core part of the export-growth model, especially in the cases of Taiwan, South Korea, and Singapore, was a strong commitment by the government to build up the international competitiveness of domestic industry.[28] While these countries took the basic industries they started with and boosted production, there was also a complete transformation of the economic structure as they revamped old sectors and branched into new with the government and private sector cooperation.[29] Moreover, shifts were made in Japan and Korea from light to heavy industry while the smaller economies, Hong Kong and Singapore, added new product lines, all of which deepened and broadened their economic bases. Equally significant, the socioeconomic dimension was not entirely forgotten. As Jon Woronoff noted: "Although often neglected and insufficiently stressed, they have all brought very definite benefits to their people. As incomes rose, it was possible to buy more and better food, clothing and housing (the latter sometimes subsidized by the state). The educational systems improved as did the hospitals and social facilities."[30]

Another crucial factor in the little tigers' success was their ability to manipulate foreign trade and investment to their advantage. Stated in another fashion, each of these states was guided by a trading class that willingly made use of new technology (often imported), which allowed them to maintain considerable flexibility when confronted by changing international economic situations.[31] This also meant that the technological level of the tigers rose, transforming them from copiers into innovators.[32]

While Singapore, Hong Kong, Taiwan, and South Korea embarked upon a turbo-charged, export-driven economic growth path, a number of developing countries (especially in Latin America as with Brazil and the Middle East as with Turkey) gravitated to the import substitution model. The 1970s proved to be a testing ground for the trading state. While the Asian trading states *began* with import-substitution programs, it is critical to note three major differences between them and their counterparts elsewhere in the developing world. First, in the cases of Taiwan and South Korea rural reform stimulated a rapid increase in agricultural productivity, the surplus of which increased productivity that was in turn channelled into the industrial sector. Secondly, Taiwan and South Korea enjoyed substantial foreign aid that included capital, equipment, commodities, and advisors. Third and finally, in all four countries, education was an important element in moving away from unskilled labor to more skilled production.

The tigers' success flew in the face of the Marxist-Leninist school and the dependency theorists. Alice H. Amsden commented: "For if, in fact, participation in foreign trade and the presence of foreign investment are useful categories for understanding the failure of Third World countries to develop, then a case like Taiwan, where development has succeeded, should be explicable by the absence of foreign trade and investment. But this is clearly incorrect."[33] She clarified this further by noting that Taiwan was popular for foreign investment and open to international trade. In the early 1990s, Taiwan has over $80 billion in foreign exchange reserves, much more than most including many developed countries.

New Capitalism and the End of Communism

While Marxism-Leninism was increasingly discredited in the late 1970s and throughout the 1980s, the Keynesian model of capitalist development came under attack in the industrialized world. As John Oliver Wilson noted: "Keynes was blamed for the decline in savings in the United States, and for the low level of new investment. The tendency of governments to operate with large budget deficits year after year was laid on his doorstep. He was charged with leading the global economy down the path of stagflation."[34] The search was on in the early 1980s for a more dynamic path of capitalistic development. This was observable by the rise of "conservative" economic policies under Prime Minister Margaret Thatcher in the United Kingdom and Prime Minister Brian Mulroney in Canada.

In the United States, the administration of Ronald Reagan entered office in 1980 and embarked upon what was defined as a supply side economic experiment that emphasized deregulation and was partially defined by the Laffer Curve theory. Simply stated, it was thought that if enough supply became available, disincentives were levelled against savings by government policies, and burdensome government regulations were removed, the magic of the marketplace would be freed and self-sustaining economic growth would result. Indeed, in the United States, the Reagan administration presided over a recovery from a deep recession and one of the longest peacetime periods of economic growth.

The then self-evident success of the supply-side model was hoisted as the correct path of development for the developing world, especially in countries where rigid statist policies and import substitution had miserably failed.

The "new capitalism" of the 1980s emphasized "democratic-capitalism." Harkening back to neoclassical economic thinking, it provided the U.S. a stronger model to hold up to the developing world. Michael Novak, a leading voice in this school, defined democratic capitalism as "three systems in one: a predominantly market economy; a polity respectful of the rights of the individual to life, liberty, and the pursuit of happiness; and a system of cultural institutions moved by the ideals of liberty and justice for all."[35] He further argued that the link between a democratic political system and a market economy was not an accident of history, but that "political democracy is compatible in practice only with a market economy" because "both systems nourish and are best nourished by a pluralistic liberal culture."[36]

The key elements of Novak's thinking, which were avidly used by the Reagan administrations in the 1980s in addressing the developing world and the Eastern bloc, were that the power of the state was to be limited to defense and to liberate the energies of individuals. Clearly, calls to liberate citizens from crippling taxation, heavy bureaucracy, and dreary regulations of state had an echo in many developing countries. As Martin Anderson, President Reagan's chief domestic and economic policy adviser in the White House senior staff, commented in the excitement of falling communism at the end of the 1980s, "And throughout the world the philosophy of socialism and communism have been discredited, as country after country watches the glittering dream of socialist theory turn to ash."[37]

The Eastern bloc's unravelling in the late 1980s vindicated many of the criticisms levelled against the Marxist-Leninist model and appeared to reinforce the tenets of new capitalism being advocated by Novak, Anderson, and others. As early as the late 1970s, the People's Republic of China had opted to introduce more capitalist economic practices. Post-Maoist reformers looked back on their country's economic experience and argued that the Revolution's Great Helmsman had gone too far in making a virtue of self-reliance.[38] Under the leadership of Deng Xiaoping, China embarked upon the Four Modernizations plan that deemphasized central planning, encouraged local initiative, and opened the door to the outside world. By the mid-1980s, China's economic policy was less red and more pink as the economy opened up to foreign trade and investment. China's strict adherence to Marxism-Leninism-Maoism, however, was unglued by mid-decade. In its place, a more capitalistic model was restructuring the Chinese economy. As the decade came to a close, considerable strides were made in augmented productivity, distribution, and standard of living. There were also considerable problems—rising inflation, corruption, regional differences, and in June 1989, political demands for a more open political system, which resulted in an authoritarian crackdown in Tiananman Square.

In Eastern Europe, Hungary, Czechoslovakia, and Poland were practicing economic policies that were increasingly capitalistic. Other socialist economies such as Cuba, Vietnam, and Romania that remained locked in central planning and rigid production codes and refused to allow information to flow freely reached a development cul-de-sac in the 1980s. Nowhere was this more evident than in the Soviet Union. By the mid-1980s, the Soviet economy was under considerable strain due to the many inefficiencies inherent in a nonmarket system as well as the costs of a protracted and unpopular war in Afghanistan and an even more onerous arms race with the United States. The grave condition facing Mikhail Gorbachev and his government in 1986 resulted in *perestroika,* an effort to open the economy.

Gorbachev's efforts to reform the Soviet economy piecemeal, however, were to fail. In the late 1980s Soviet economists were forced to abandon their traditional theory of central planning and move along a bumpy road toward the market. As the Union of Soviet Socialist Republics became unglued in the 1989–1991 period, a more neoclassical school of economists emerged, pushing the government to open the economy to world markets as well as for a less pervasive state. This liberal school, in fact, concurred with the neoclassical school as to the state's role. According to one analyst, the ideal would be a state that "would concede part of its power to markets and political civil society, one that would, furthermore, as has been increasingly argued, also privatize part of its property."[39]

By 1 January 1992 the Soviet Union had voluntarily voted itself out of existence and was replaced by the Commonwealth of Independent States (CIS). This signalled that the development debate between capitalism and communism has ended with the overwhelming victory of the former. As

France's Bernard-Henri Lévy accurately foresaw in 1977, "Like all frauds, this one [Marxism-Leninism] hardly stands up now to the only meaningful test: concrete history and its cruellest lessons."[40] This was reflected in the fall of the Berlin Wall, the bloody end of the communist regime in Romania, the arrest of T. Zhikov, Bulgaria's long-time communist strongman, and elections throughout the region, even including Albania. The defeat of communism and socialism in the development debate was reflected by the title of June Wanniski's article in the Spring 1992 issue of *Foreign Affairs*: "The Future of Russian Capitalism."[41]

The Post-Cold War Development Debate: Contending with the Trading State

In the trading states such as Japan the supply side economic experiment was watched with interest. While there was satisfaction with the failure of communism and the demise of rigid, centrally planned economies, there was little inclination for government to abandon the priority attention given to flexible industrial policy. To the leaderships of trading states the question was simple: Why should they give up the economic structure that enhances their national competitiveness? This viewpoint led to protracted and ongoing discussions between the United States and Japan and other Asian tigers about trade policies.

As the threat of the cold war diminished, the success of the tigers came under greater scrutiny. The tigers' trading state approach is in contrast to the more neoclassical system in the United States and Western Europe in which the government refrains from intervention in the marketplace except in limited cases, such as antitrust regulation and the protection of consumer rights (i.e., fair lending practices in banking). While the trading states are part of the capitalist family, they differ from the neoclassical states because their governments intervene actively in the economy in order to guide or promote particular substantive goals (e.g., full employment, export competitiveness, energy self-sufficiency).[42]

The tigers in the late 1980s were dynamic trading states, guided by an activist state philosophy. They increasingly challenged older, neoclassical states, such as the United States and much of Europe for market space. Within the trading state club even Japan found itself coming under pressure in certain markets, especially from Korea. Equally significant for the growing international weight of the Tigers was the emergence of a new wave of trading states in the cases of Malaysia, Indonesia, Thailand, and to a lesser extent, Chile and Mexico.[43]

The demise of communism meant that the rise of the Tigers also had an international power element. By the late 1980s, after the Reagan administration ended and the Bush administration began (1988–1992), the United States' economy swung into a deep recession. The trading states still had populations with substantial savings and with each devaluation of the dollar, it appeared that Japan's power grew, almost overshadowing the United States. While the United States, following the economic principles of Adam Smith, waited for the "invisible hand of the market" to correct imbalances of global production and distribution, the trading states continued to be guided by the "visible hand of the market." As Steven Schlossstein commented: "Needless to say, the Japanese do not subscribe to Adam Smith and never have.... A popular joke, which one hears frequently in Tokyo these days, captures the essence of what the Japanese consider to be Adam Smith's shortcomings: How many neoclassical economists does it take to change a light bulb? The answer: None; they simply sit and wait for the invisible hand of the market to do it for them."[44] The disputes and differences between the U.S. form of capitalism and that of Japan and the little tigers has not been lost on the next round of tigers that are in the process of emerging. Although Japan swung into a steep recession in the early 1990s, it has continued to post substantial bilateral trade surpluses with the U.S. Trade continues to be a difficult issue between Tokyo and Washington, reflecting that the next round of global competition is between two different ideas of capitalism.

Economic Versus Political Reform: What Comes First?

No discussion about development is complete without considering the political aspects of the trials and tribulations of political development and its relationship to economic development. This is of particular significance in examining the new round of competition within the larger realm of capitalist models. Added to this, political and economic development will be considerably complicated in the 1990s and 2000s by a plethora of ethnic, racial, and religious tensions. These fissures fly in the face of the idea of an interdependent world linked by common values of open markets and political systems. Moreover, the goal of a democratic and capitalist world, preferred by the West, will find a difficult road ahead as attested in a substantial literature.

Samuel P. Huntington's *Political Order in a Changing Society* (1968) was one of the earliest and most influential studies on political development. The key point of his study is that the process of development, both economically and politically, is potentially more disruptive than order oriented. Consequently, as economic change gains momentum, political institutions come under considerable strain, often caused by new demands from the population. In many cases, as history has demonstrated, those political institutions have been wanting and some suffer from "political decay," ultimately collapsing.[45]

The concerns raised in Huntington's book still have validity in today's world and, in particular, to this study. Social scientists continue to grapple with the problem as reflected by Larry Diamond, Juan J. Linz, and Seymour Martin Lipset: "Among the most important dimensions of the state-society relationships is the strong tendency we find for state dominance over the economy and society to undermine democratic politics in developing countries."[46]

Democracy, in the Western liberal democratic tradition, appeared to have broken out almost everywhere in the late 1980s (with the notable exception of the Middle East), and then contracted (Haiti, Peru, and a number of African states). It is now struggling to stay afloat in many countries, ranging from Russia and Romania to Peru and Brazil, and it can be argued that political institutions in all those countries are hard-pressed to meet new demands. Events in Guatemala, Nigeria, and Burma reflect the difficult path ahead in the transition from authoritarian regimes to elective governments. Even armed by the mandate provided by election by popular vote, leaders find the challenges raised by economic and political development to be tough—as shown by the ousters of presidents Fernando Collor in Brazil and Carlos Andrés Pérez in Venezuela. Therefore, some of the most difficult debates facing those involved in the development game in the coming years are: What comes first, economic or political development? Does political development have to follow the lines laid out by liberal Western democracies?

The argument can be made that democracy in the developing world remains precarious and incomplete at best. Democracy or democratically inclined forces face a myriad of threats. However, corruption, the entrenched power of inward-looking elites, a lack of consensus about economic direction, and poverty are probably the most difficult. The prevalence of poverty among large swathes of Africa, Latin America, Asia, and the Middle East will continue to jeopardize political stability and growth, while fostering populist rhetoric or holding the door open to those bold enough to seize power by armed force. The absence of opportunities, beyond criminal endeavors (such as drug or arms smuggling) is particularly crippling. As former U.S. State Department official Elliott Abrams commented, "The system still seems designed to stifle initiative and competition, and to ensure that children grow up to be as poor (or as rich) as their parents."[47] Although this charge was leveled at Latin America, its resonance is not limited to that region as proved by the examples of Nigeria and the Philippines (especially under Ferdinand Marcos).

Capitalism, or the "magic of the market," is often waved about as a banner that will eventually sweep away all that ails a society. Yet, a lack of controls can be as dangerous to a marketizing society as too many restrictions. In a system in which a political leader and his cronies are the only ones to benefit, economic reforms will ultimately fail and capitalism assumes negative connotations with the majority of the population. Examples of such "crony capitalism" include the Philippines under the Marcos regime and a number of the military governments in Nigeria.

While crony capitalism does indeed describe the style of "economic reform" in a number of developing countries, it is also true that economic reform that advances marketization has been surprisingly popular at the polls if only because of the lack of other viable alternatives. Grumbling and rhetoric aside, voters from Argentina to South Korea are generally supporting reformers rather than socialists. The future, however, is less clear. Should it become manifest that opportunities do not exist for the general population due to "crony capitalism" or racial discrimination, economic reform will indeed be doomed.

How then to break the hold of the (often corrupt) elites, who oppose or hinder change, and to foster lasting and meaningful economic and political reform? The route to success employed by the Asian tigers may be one way. A 1993 World Bank study of these "economic miracles" tried to help economists learn from the extraordinary gains of East Asia, a region where per capita income nearly quadrupled over the past quarter century. The study found that policies followed by different countries throughout the region were successful, despite a marked divergence—Hong Kong's laissez-faire brand of capitalism versus the more activist state roles assumed in South Korea and Singapore. The World Bank study suggests that the tiger economies used a mixture of market incentives and state interference in key areas. Privatization, moreover, has been a gradual process. Remarkable productivity growth only partly reflects market-oriented policies. In addition, all of the high fliers intervened selectively to promote/protect particular industries, or to forcibly shift resources from consumption to investment-directed growth.

These tentative conclusions, if accurate, create a dilemma for those advising the new crop of would-be tigers. Does the conventional prescription of speedy deregulation and privatization really make sense given the Asian lessons? Or should they be promoting East Asia's slower, more pragmatically interventionist path to economic development? Ample grounds for caution are available (even though government activism outside East Asia has often had dismal results). One key lesson, though, is that developing economies apparently do need the government's guiding hand; development cannot be left to the private sector alone.[48] The trick is create the correct balance. Too much government smothers initiative; too little allows the rise of crony capitalism, which provides abundant ammunition for radical leftists who are down but not entirely out.

Directly related to the question of too much or too little government is the issue of open versus closed politics, especially in the context of competing capitalist models. While Communist totalitarianism is out as a challenge to democratic-capitalism, there are some that feel that "soft" or "soft shell" authoritarianism is in. According to Francis Fukuyama, political scientist Robert Scalapino's soft authoritarianism is becoming increasingly popular with the Asian tigers as a "potential competitor to Western liberal democracy." Fukuyama indicates that such a model has two distinguishing components: a market-oriented economic system with "a kind of paternalistic authoritarianism that persuades rather than coerces," and a value system that "emphasizes conformity to group interests over individual rights."[49]

Soft authoritarianism in the Asian milieu reflects an influence of Confucian values. According to Denny Roy, lecturer in political science at the National University of Singapore, these values "champion order, a strong but moral state, and the needs of society as a whole over personal freedoms and limitations on government."[50]

While capitalism has won out against communism and socialism, the new contest within capitalism includes proponents of soft authoritarianism both inside and outside Asia. They believe that this model offers a better framework for political and economic development, especially as it provides and legitimizes strong leadership in a process that Huntington indicated is not conducive to political order. Moreover, proponents of soft authoritarianism argue that the fruits of Western liberal democracy may be overrated. As Heng Chiang Meng of the ruling People's Action Party in Singapore commented of U.S. democracy: "To walk the streets with reasonable safety is the most basic of civil liberties. Yet millions of Americans dare not step out at night and some scarcely dare to venture out by day.... In every American city pupils are assaulted daily and some killed."[51]

It is important to underscore that soft authoritarianism is also observed by some analysts as part of a process in reaching greater political and economic liberalization. The emphasis is on economic reform first followed by political reform. This process is illustrated by, in Asia, the cases of Taiwan and South Korea and, in Latin America, Chile and perhaps Mexico. Roy summarizes the issue of competing capitalist political models thusly: "Soft authoritarianism may be transitory, a stepping stone to a form of government more resilient against the pressures faced by modern nation-states. Soft authoritarianism may yet succeed Western liberalism as the new hegemonic discourse, or it may wind up on the 'rubbish heap of history' with no appeal outside of insecure political parties whose interests it directly serves."[52]

Ultimately, the answer to the question of what should come first, political or economic development, is that each country is unique and its government and should decide how to approach development by playing to its greatest strengths. This means that if a country already has a democratic political system, it by no means should adopt any authoritarian regime. One the other hand, a less open society will find that its political development may flounder if it seeks to make an immediate conversion to an elective system without the establishment of democratic ground rules and without the economic means to maintain progressive development that benefits a wider range of society. While the authors share a marked preference for democratic political systems, they do recognize that countries such as Taiwan and South Korea have followed more evolutionary paths on the political front and are, at their own pace, establishing more open political systems.

Conclusion

The developing world ended the 1980s conducting a range of economic experiments, most of which were capitalist oriented and influenced by the ideas of free trade, comparative advantage, and, in some cases, supply side economics. The experiment of supply side economics in the United States clearly influenced the thinking of policymakers in countries as diverse as Mexico, Ghana, and Sri Lanka. The fact that the Soviet model of development was such a failure echoed back to the ideas of Adam Smith that favored an individual's freedom of choice—especially in the economic field. At the same time, the trading states' economic model gained attention and many attributes of it were copied. Moreover, the trading/developmental state poses a model that stands in contrast to the neoclassical model of open economies and free trade. This is even more the case if the soft authoritarian model is added to the discussion. In a sense, the demise of communism has set the stage for the next round in the development debate—between the trading/developmental states' mode of capitalism and that of the neoclassical economies, calling for free trade, or, from the political perspective, between the soft authoritarian and more open political systems.

For the newly emerging capitalist states in Eastern Europe and in what was the Soviet Union, the neoclassical model of development is regarded as the correct path, especially in light of the sweeping failure of communism. In Latin America, the statist policies influenced by the dependency school were jettisoned in a number of cases, such as Chile and Mexico. What model of capitalism will the states in Central Asia, Africa, and Latin America ultimately adopt? The answer is likely to be complicated, but will involve an evolution to more open economies and polities.

Economic development is a difficult process. Considering that economic policies have obvious effects on society, they become political issues that are often fought over. By the advent of the last decade of the twentieth century, the development debate between the U.S.-led West and the Soviet-led East was over. The former had won—overwhelmingly. New debates within the capitalist system have already arisen in its stead. The next phase of the development debate will be tested by the efforts of developing nations to find the best route to add themselves to the ranks of tigers and to avoid the company of elephants.

Notes

1. Peter Donaldson, *Economics of the Real World* (London: British Broadcasting Corporation and Penguin Books, 1973, second edition 1978), 11.
2. George Dalton, *Economic Systems and Society: Capitalism, Communism and the Third World* (London: Penguin Books, 1978), 28–29.
3. Adam Smith, *The Wealth of Nations*, as quoted in Dalton, *Economic Systems and Society*, 263.
4. David Fusfeld, *The Age of the Economist* (Glenview, Ill.: Scott, Foresman and Company, 1977), 36–37.
5. It is important to clarify that many of these problems, such as child labor, existed before in agrarian society, but the industrial revolution caused greater awareness. There was a diffusion downward to those who had become rich in trade of the sense of noblesse oblige, which were transformed into Victorian ideals of philanthropy. Equally significant, the rise of literacy, use of photographs, and newspapers carried the images of the day. One reaction to the changes in English society was Charles Dickens' *Hard Times*.
6. Pekka Setula, *Economic Thought and Economic Reform in the Soviet Union* (Cambridge: Cambridge University Press, 1991), 1.
7. John Oliver Wilson noted:

   > While the other major industrialized countries were not so open in their acceptance of Keynesian views, they followed many of the same principles. The French were reluctant to embrace the theories of an English-speaking economist, yet their own economists integrated many of the Keynesian policy prescriptions into French thinking. The Japanese did not need an economic theory to convince them of the importance of strong government leadership in the economy; they were well embarked upon such a course for their own unique reasons. And the Germans, while cautious to avoid the excesses of Keynesianism for fear of inciting inflation, nevertheless wrote Keynes's basic tenets into law in 1967 in their Stability and Growth Law for the Economy.

   The Power Economy: Building an Economy That Works (Boston: Little, Brown and Company, 1985), 15.
8. Donaldson, *Economics of the Real World*, 83.
9. Nathan Rosenberg and L. E. Birdzell, Jr., *How the West Grew Rich: The Economic Transformation of the Industrial World* (New York: Basic Books, Inc., 1986), 330.
10. Alec Nove, *An Economic History of the U.S.S.R.* (Harmondsworth: Penguin Books, 1975), 265.
11. See Choh-Ming Li, "China's Industrial Development, 1958–63," in Roderick MacFarquhar, ed., *China Under Mao: Politics Takes Command* (Cambridge, Mass.: The MIT Press, 1972), 175–211.
12. Benjamin Schwartz, "Modernization and the Maoist Vision—Some Reflections on Chinese Communist Goals," in MacFarquhar, ed., *China under Mao*, 15.
13. An example of this entrepreneurial elite was noted by Rostow:

> A requirement for take-off is, therefore, a class of farmers willing and able to respond to the responsibilities opened up for them by new techniques, land-holding arrangements, transport facilities, and forms of market and credit organization. A small purposeful elite can go a long way in initiating economic growth; but, especially in agriculture (and to some extent in the industrial working force), a wider-based revolution in outlook must come about.

W. W. Rostow, *The Stages of Economic Growth: A Non-Communist Manifesto* (Cambridge: Cambridge University Press, 1961), 51–52.

14. Michael Todaro, *Economic Development in the Third World* (New York: Longman, 1981), 61.
15. Nigel Harris, *The End of the Third World: Newly Industrializing Countries and the Decline of an Ideology* (Harmondsworth: Penguin Books, 1986), 7.
16. Todaro, *Economic Development in the Third World,* 68.
17. Joan Robinson, "The Second Crisis of Economic Theory," *American Economics Association Papers and Proceedings* (May 1972), 8.
18. Donaldson, *Economics of the Real World,* 234–35.
19. As Donaldson noted, "But quite apart from the fact that the growth performance of the rich economies was even better during these years, and that a good deal of increased output in poor countries was swallowed up by rapidly increasing population, the nonsense of equating growth with progress is even more patently obvious in the case of underdeveloped economies than it is in the affluent West." *Economics of the Real World,* 236.
20. André Gunder Frank, *Lumpenbourgeosie: Lumpendevelopment; Dependence, Class and Politics in Latin America* (New York: Monthly Review Press, 1972), 15. Other dependency school members include James Petras, *Politics & Social Structure in Latin America* (New York: Monthly Review Press, 1970); Paul Baran, *The Political Economy of Growth* (New York: Monthly Review Press, 1962); Immanuel Wallerstein, "Semi-peripheral Countries and the Contemporary World Crisis," in *Theory and Society 3,* 4 (1976): 461–83; and Fernando Henrique Cardoso and Enzo Faletto, *Dependency and Development in Latin America* (Berkeley, Calif.: University of California Press, 1979). A useful review of early dependency thinking is Gabriel Palma, "'Dependency' A Formal Theory of Underdevelopment or a Methodology for the Analysis of Concrete Situations of Underdevelopment?" *World Development 6* (November 1978): 881–924. Another useful review is Stephan Haggard, *Pathways from the Periphery: The Politics of Growth in the Newly Industrializing Countries* (Ithaca, N.Y.: Cornell University Press, 1990), see chap. 1.
21. Frank, *Lumpenbourgeosie,* 145.
22. Rosenberg and Birdzell, *How the West Grew Rich,* 331.
23. Harris, *The End of the Third World,* 7.
24. Richard Rosecrance, *The Rise of the Trading State: Commerce and Conquest in the Modern World* (New York: Basic Books, Inc., Publishers, 1986), ix.
25. World Bank, *World Development Report 1991* (New York: Oxford University Press, 1991), 204–5.
26. Ibid.
27. Joan Robinson, *Aspects of Development and Underdevelopment* (Cambridge: Cambridge University Press, 1979), 109.
28. Robert Wade, *Governing the Market: Economic Theory and the Role of Government in East Asian Industrialization* (Princeton: Princeton University Press, 1990), 7. Wade asserts this is the case for Japan, Taiwan, and South Korea. The authors have added Singapore to this assertion.

29. Jon Woronoff, *Asia's 'Miracle' Economies* (Armonk, N.Y.: M. E. Sharpe, Inc., 1986), 10.
30. Ibid., 10.
31. See Rosenberg and Birdzell, *How the West Grew Rich,* for a variation on this theory.
32. Ibid.
33. Alice H. Amsden, "Taiwan's Economic History: A Case of Etatisme and a Challenge to Dependency Theory," in Robert H. Bates, ed., *Toward a Political Economy of Development* (Berkeley: University of California Press, 1988): 171–72.
34. Wilson, *The Power Economy* (Boston: Little, Brown and Company, 1985), 18.
35. Michael Novak, *The Spirit of Democratic Capitalism* (New York: A Touchstone Book, 1982), 14.
36. Ibid.
37. Martin Anderson, *Revolution* (New York: Harcourt Brace Jovanovich, Publishers, 1988), 8.
38. John Gittings, *China Changes Face: The Road from Revolution 1949–1989* (Oxford: Oxford University Press, 1990), 227.
39. Sutela, *Economic Thought and Economic Reform in the Soviet Union,* 160.
40. Bernard-Henri Lévy, *Barbarism with a Human Face* (New York: Harper & Row Publishers, 1977), 5.
41. See June Wanniski, "The Future of Russian Capitalism," *Foreign Affairs* (Spring 1992): 17–25.
42. Steve Chan, *East Asian Dynamism: Growth, Order, and Security in the Pacific Region* (Boulder, Colo.: Westview Press, 1990), 47–48.
43. For a discussion about the development of new trading states and the significance of trade to Malaysia, see Mohamed Ariff's *The Malaysian Economy: Pacific Connections* (Singapore: Oxford University Press, 1991).
44. Steven Schlosstein, *The End of the American Century* (New York: Congdon and Weed, Inc., 1989), 29.
45. Samuel P. Huntington, *Political Order in a Changing Society* (New Haven, Conn.: Yale University Press, 1968).
46. Larry Diamond, Juan J. Linz, and Seymour Martin Lipset, *Politics in Developing Countries: Comparing Experiences with Democracy* (Boulder, Colo.: Lynne Rienner Publishers, 1990), 24.
47. Elliott Abrams, "How To Avoid the Return of Latin Populism," *The Wall Street Journal* (21 May 1993): A24.
48. Michael Prowse, "Miracles Beyond the Free Market," *The Financial Times* (26 April 1993).
49. Francis Fukuyama, "Asia's Soft Authoritarian Alternative," *New Perspectives Quarterly* 9, 2 (Spring 1992): 60–61.
50. Denis Roy, "Singapore, China, and the 'Soft Authoritarian' Challenge," *Asian Survey xxxiv,* 3 (March 1994): 231.
51. Cao Yunhua, "Singapore as a Role Model," *Sunday Times* (12 April 1992): 22; *Straits Times* (28 May 1988): 20.
52. Roy, "Singapore, China, and the 'Soft Authoritarian' Challenge," 242.

References

Anderson, Martin. *Revolution.* New York: Harcourt Brace Jovanovich, Publishers, 1988.

Ariff, Mohamed. *The Malaysian Economy: Pacific Connections.* Singapore: Oxford University Press, 1991.

Bannock, G., R. E. Baxter, and R. Rees. *The Penguin Dictionary of Economics.* Harmondsworth: Penguin, 1972.

Baran, Paul. *The Political Economy of Growth.* New York: Monthly Review Press.

Bates, Robert H., ed. *Toward a Political Economy of Development.* Berkeley: University of California Press, 1988.

Cardoso, Fernando Henrique and Enzo Faletto. *Dependency and Development in Latin America.* Berkeley: University of California Press, 1979.

Chan, Steve. *East Asian Dynamism: Growth, Order, and Security in the Pacific Region.* Boulder, Colo.: Westview Press, 1990.

Dalton, George. *Economic Systems and Society: Capitalism, Communism and the Third World.* London: Penguin Books, 1978.

Donaldson, Peter. *Economics of the Real World.* London: British Broadcasting Corporation and Penguin Books, 1973, second edition 1978.

Diamond, Larry, Juan J. Linz, and Seymour Martin Lipset. *Politics in Developing Countries: Comparing Experiences with Democracy.* Boulder, Colo.: Lynne Rienner Publishers, 1990.

Frank, André Gunder. *Lumpenbourgeosie: Lumpendevelopment; Dependence, Class and Politics in Latin America.* New York: Monthly Review Press, 1972.

Fry, Maxwell. *Money, Interest, and Banking in Economic Development.* Baltimore, Md.: The Johns Hopkins University Press, 1988.

Fukuyama, Francis. "Asia's Soft Authoritarian Alternative." *New Perspectives Quarterly 9,* 2 (Spring 1992).

Fusfeld, David. *The Age of the Economist.* Glenview, Ill.: Scott, Foresman and Company, 1977, third edition.

Gastmann, Albert L. *Historical Dictionary of the French and Netherlands Antilles.* Metuchen, N.J.: The Scarecrow Press, Inc., 1978.

Gittings, John. *China Changes Face: The Road from Revolution 1949–1989.* Oxford: Oxford University Press, 1990.

Haggard, Stephan. *Pathways from the Periphery: The Politics of Growth in the Newly Industrializing Countries.* Ithaca, N.Y.: Cornell University Press, 1990.

Harris, Nigel. *The End of the Third World: Newly Industrializing Countries and the Decline of an Ideology.* Harmondsworth: Penguin Books, 1986.

Huntington, Samuel P. *Politcal Order in a Changing Society.* New Haven, Conn.: Yale University Press, 1968.

Keynes, John Maynard. *The General Theory of Employment, Interest, and Money.*

Lévy, Bernard-Henri. *Barbarism with a Human Face.* New York: Harper & Row Publishers, 1977.

MacFarquhar, Roderick, ed. *China Under Mao: Politics Takes Command.* Cambridge, Mass.: The MIT Press, 1972.

McDonald, Lee Cameron. *Western Political Theory: Part 2, From Machiavelli to Burke.* New York: Harcourt Brace Javonovich, Inc., 1968.

Novak, Michael. *The Spirit of Democratic Capitalism.* New York: A Touchstone Book, 1982.

Nove, Alec. *An Economic History of the U.S.S.R.* Harmondsworth: Penguin Books, 1975.

Palma, Gabriel. "Dependency: A Formal Theory of Underdevelopment or a Methodology for the Analysis of Concrete Situations of Underdevelopment?" *World Development 6* (November 1978).

Petras, James. *Politics & Social Structure in Latin America.* New York: Monthly Review Press, 1970.

Polanyi, Karl. *The Great Transformation.* Boston: Beacon Press, 1957.

Robinson, Joan. *Aspects of Development and Underdevelopment.* Cambridge: Cambridge University Press, 1979.

______. "The Second Crisis of Economic Theory." *American Economics Association Papers and Proceedings* (May 1972).

Rosecrance, Richard. *The Rise of the Trading State: Commerce and Conquest in the Modern World.* New York: Basic Books, Inc., 1986.

Rosenberg, Nathan and L. E. Birdzell, Jr. *How the West Grew Rich: The Economic Transformation of the Industrial World.* New York: Basic Books, Inc., 1986.

Rostow, W. W. *The Stages of Economic Growth: A Non-Communist Manifesto.* Cambridge: Cambridge University Press, 1961.

Roy, Denny. "Singapore, China, and the 'Soft Authoritarian' Challenge." *Asian Survey xxxiv,* 3 (March 1994).

Schlossstein, Steven. *The End of the American Century.* New York: Congdon and Weed, Inc., 1989.

Setula, Pekka. *Economic Thought and Economic Reform in the Soviet Union.* Cambridge: Cambridge University Press, 1991.

Todaro, Michael. *Economic Development in the Third World.* New York: Longman, 1981.

Wade, Robert. *Governing the Market: Economic Theory and the Role of Government in East Asian Industrialization.* Princeton: Princeton University Press, 1990.

Wallerstein, Immanuel. "Semi-peripheral Countries and the Contemporary World Crisis." *Theory and Society 3,* 4 (1976).

Wanniski, June. "The Future of Russian Capitalism." *Foreign Affairs* (Spring 1992).

Wilson, John Oliver. *The Power Economy: Building an Economy That Works.* Boston: Little, Brown and Company, 1985.

World Bank. *World Development Report.* New York: Oxford University Press, Annual.

Woronoff, Jon. *Asia's 'Miracle' Economies.* Armonk, N.Y.: M. E. Sharpe, Inc., 1986.

Yunhua, Cao. "Singapore as a Role Model." *Sunday Times* (12 April 1992); *Straits Times* (28 May 1988).

Part II

New Tigers Burning Bright

3

A Tiger that Tangos: Argentina

The long and exaggerated decline of Argentina in this century is legendary.[1] Its social, economic, and political development in 1930 compared favorably to that of Canada, Australia, and much of southern Europe. Blessed with some of the world's most fertile farmland, ample energy resources, and a well-educated workforce, it is truly remarkable that Argentina has not parlayed those assets into economic or political stability. According to one observer, "virtually all economists agree that Argentina has all the ingredients for a sound and stable economy."[2] Yet, as another economist points out: "By the early 1980s Argentina was a quintessential example of a country that had squandered its resources, failed to fulfill the economic aspirations of its people and marginalized itself internationally. It had a wealth of mineral resources, including hydrocarbons and fertile soil, and a strategic location in the southern Atlantic, but the record of the country's leadership in harnessing what the country offers was woeful."[3] By the end of the 1980s the country had slipped, seemingly irrevocably, from the First World to the Third World. Some five years later, a turnaround had begun, but serious doubts remain as to the long-term viability of the reforms.

Argentina suffered a devastating series of public humiliations during the 1980s from which many observers thought it would never recover. The country bumbled from its military defeat by British troops in the Falkland Islands in 1982 to a dubious position as the basket case economy of Latin America, surpassing even Brazil and Peru for this distinction. The decade of the 1980s was one of massive inflation and currency devaluation, of zero to negative growth in the economy, and plunging living standards. By 1989 the data told a sorry but expressive tale: inflation for the year stood at a record-breaking 3079 percent, the economy contracted by 6.2 percent, and the budget deficit was running at more than 10 percent of GDP. The economy was characterized by a bankrupt state treasury; aggressive labor unions locked in mortal combat with top-heavy, inefficient business cartels; and a huge, hopelessly bloated public sector. Above all was the constant stench of endemic corruption and mismanagement, which laid a miasma of depression over the entire country.

It is nothing short of remarkable, then, that we are now identifying Argentina as the foremost tiger of the 1990s in Latin America. Under the unlikely leadership of Peronist President Carlos Menem, the country has been reborn as a bastion of free-market economic policies and private sector enterprise. The economic miracle taking place under Menem and his Economy Minister Domingo Cavallo is setting the stage for a genuine and lasting takeoff that should encompass the 1990s and beyond despite the difficult international market in early 1995. Coupled with the structural strengths of the Argentine economy (which were successfully trampled by decades of gross mismanagement), the reforms now taking place should enable Argentina to advance into the modern industrialized world.

Myriad challenges remain, of course. Argentina has not yet permanently tamed corruption or inflation, as we will discuss. Also, the importance of personalities, Menem's and Cavallo's, in particular, highlights the fragility of the reform process still in its infancy. Moreover, investment is still well below the levels needed to sustain 6 percent plus growth rates over the next decade. Credible and ambitious economic policies will be required to attract the return of flight capital and build domestic savings levels to the point that they can support much higher investment expenditures. Many companies are still antiquated by international standards, and the labor market is still painfully rigid. Blips on the privatization process have appeared, and there are deepening concerns as to how the government's finances can be stabilized once it can no longer depend on privatization receipts. Social tensions are mounting, and worse conflicts lie ahead as the industrial shakeout gathers steam. The question of sustainability depends most crucially on the government's ability to maintain an overvalued exchange rate in the face of ballooning current account deficits and dwindling export competitiveness.

For the many businesses and investors contemplating Argentina, if there were no risk, there would be no return. In their judgment, Argentina is worth the risk. According to the International Monetary Fund, foreign direct investment has shifted from a small net outflow in 1987 to net inflows of $4.2 billion in 1992 and $6.2 billion in 1993.[4] Indeed, Argentina always possessed the strengths and skills necessary to take its place among world-class economies. The distinction between a genuinely poor country and a poorly managed one is particularly useful for Argentina. The self-destructiveness of economic policymakers there since World War II is nothing short of remarkable, transforming an economy with living standards at Canadian levels in the prewar years (Argentina was once one of the six richest countries in the world) to one firmly mired in the Third World by 1990. Pragmatic and far-reaching policy changes under Menem have the potential for turning Argentina into the OECD-level economy that it always should have been.

Advantage: Technology

As living standards have deteriorated drastically since World War II and chronic economic crises have taken a heavy toll, Argentina has come to suffer from the curse of inadequate social and educational spending. The economic disaster of the 1980s produced a virtual collapse in health care, steady increases in infant mortality, and minimal official response. One study found that more than 30 percent of the population are poor and malnourished while half lack adequate housing; an outbreak of cholera in 1992–1993 highlighted woeful health and sanitary conditions.[5] Obviously, budgetary constraints (caused in part by excessive military demands on the government's finances) kept educational expenditures depressed.

Despite these problems, though, Argentina has fundamental strengths in this area that differentiate it sharply from its Latin American neighbors. Literacy levels are closing in on OECD standards at 92 percent and have not declined in recent years despite the economic crises. The workforce is highly urbanized and has largely moved away from agriculture and into business, especially the services sector. A substantial middle class, much larger than in other Latin American countries, provides a sizeable pool of skilled workers. The emphasis on services and business has created the need for a reasonably sophisticated computer sector relative to Brazil or Peru, and businesses are moving rapidly to improve their capabilities in this area. All in all, the presence of an urbanized, well-educated, largely middle-class workforce in Argentina should position it well for competition in the technology-driven 1990s.

Argentina's industrial sector is also actively upgrading in terms of technology after a long drought in investment. One example of this is Edesur (*Empresa Distribuidora Sur Sociedad Anonima*), an electric distribution company that provides monopoly services in the southern part of the greater Buenos Aires metropolitan region. Privatized in 1992, Edesur embarked upon a revamping of its grid and the introduction of new technology.[6] Edesur is hardly alone in overhauling its technology—the private telephone sector has also undergone the same process, helped by its international connections related to foreign equity owners. Both privatized telephone companies have pledged to invest nearly $1 billion annually to improve telephone service.[7]

Ethnic Stability

Economic decline during the 1980s helped turn a once-wealthy nation where food was traditionally plentiful and cheap, into one harboring vast pockets of extreme poverty and class-based conflict. Despite some macroeconomic progress since 1990, the social situation continues to deteriorate as shanty towns spring up around the cities and crime levels (in a country that prided itself on low crime) escalate sharply. The urgent need to control inflation and stabilize official finances has hindered the government from introducing programs to ease the hardships and provide a safety net; indeed, government austerity programs have worsened living conditions for many pensioners.

Social tensions have risen markedly in the past few years despite overall economic progress. In the most serious challenge yet to Menem's economic program, provincial riots at the end of 1993 highlighted the dramatic deterioration of living conditions in some parts of the country. The gap between rich and poor is rising sharply, while the already huge divisions between wealthy and backward regions are also mounting. Strong economic growth since 1991 is restricted almost entirely to Buenos Aires and the industrial provinces of Cordoba and Santa Fe. At the same time, Argentina used to offer generous social benefits and government services, but decades of high inflation, economic decline, and widespread corruption have left social services in tatters. As the industrial shakeout accelerates, these tensions will undoubtedly rise.

The most serious and enduring social conflict involves organized labor and employers, which includes both corporate management and the state. In the late 1940s Juan Peron reorganized labor to include the urban unskilled workers, many of whom were new immigrants or impoverished mestizo migrants from the countryside. Peron's unions and their carefully chosen leaders were often corrupt. Each non-Peronist government since the 1955 coup against Peron has tried to break his party's union stronghold, with little success. Argentina has one of the strongest union movements in the world, encompassing over 30 percent of the labor force. Even Menem has had little success in reforming the notoriously inflexible labor sector.

But while the country has been marked by class struggle and conflict, it is blessed with relative racial homogeneity. Argentina is unique in Latin America because the huge bulk (85–90 percent) of its population is of European origin, which has eliminated much of the racial tensions plaguing Peru and even Brazil. There are probably fewer than 100,000 Indians in the entire nation, while mestizos have been gradually absorbed by waves of European immigrants. Today perhaps 10 percent of the population may be classified as mestizo, and there are virtually no blacks. Although the mestizo minority (persons with white and Indian blood) is generally restricted to unskilled jobs, there are few conflicts between the mestizos and the predominantly white and Europeanized urban population. Moreover, in sharp contrast to the typical Latin American model, there is no lack of a middle class in Argentina; almost half of the population may be in this category.[8] While Argentina has a strong history of class conflict between labor and employers in this century—and a growing mass of social tensions—these have no base in racial or ethnic divides.

Economic Reform: Menem's Good Luck

When Menem took office in July, 1989, he was lucky enough to inherit an economy that had hit rock bottom. Inflation was soaring at a rate of nearly 200 percent per month; the federal budget deficit was running at 23 percent of GDP; foreign debt was projected at $64 billion by the end of the year, with arrears on the debt nearing $5 billion; real wages dropped 50 percent in the first six months of 1989; and unemployment in greater Buenos Aires was approaching 20 percent. As the state essentially went bankrupt, it set the stage for reform. After years of economic chaos, it became clear that reducing inflation, cutting state spending, privatizing state industries, and encouraging a competitive business sector was the only route to prosperity. This produced an environment that was at least moderately amenable to the reforms that Menem introduced.

While much remains to be done, economic improvements since then under Menem and Cavallo are justifiably being dubbed the "Argentine Miracle." At long last, stability appears possible after years of chronic hyperinflation and currency devaluations, spiraling foreign debt, political instability, and capital flight. Menem has introduced a series of dramatic reform and austerity programs that have accomplished important goals. The turning point came in early 1991, when Economy Minister Cavallo introduced an economic plan designed to win the confidence of domestic and foreign investors. This program put a floor under the value of the austral, prohibited the central bank from printing money not backed by gold and foreign exchange reserves, and required the government to maintain a fiscal surplus. Menem has succeeded in dramatically downsizing the role of the state by slashing tariffs, abolishing exchange restrictions and price controls, selling off large chunks of nationalized industry, and ending government interference in private sector wage negotiations. He is trying to make Argentina's economy efficient and competitive, implementing a sweeping and dramatic program to abolish most of the regulations that had hobbled private enterprise and deterred foreign investors. These policies contrast dramatically with the Peronist legacy of state protection and intervention.

The results so far have been equally dramatic. After years of stagnation the economy expanded by a spectacular 24 percent in 1991–1993 and industrial production soared to record levels. Even more astonishing, inflation dropped to 10.6 percent by 1993 and continued to fall in 1994 to 4–5 percent, while the peso remained stable and the government's budget swung into surplus. In the 1989 to 1992 period, 670,000 jobs were created, representing a 5 percent increase in total employment.[9] The government has successfully completed a $20 billion radical privatization program, which culminated with the 1993 sale of Argentina's largest company, the state oil firm YPF. Most encouraging of all for the long term, Argentina's economic policies have won glowing endorsements from its old nemesis the IMF, enabling the country to conclude a Brady Plan debt restructuring and reduction agreement with its foreign commercial bank creditors. Capital inflows (both actual and contracted) have rebounded, totalling around $8 billion in 1992 and an estimated

$16.5 billion in 1993. An optimistic Cavallo predicts that his country is on the verge of a spectacular upsurge in investment growth, expecting investment to double from under 10 percent of GDP in recent years to 20 percent in the next few years. This would be at least as great an achievement as his defeat of hyperinflation, and would insure that his market-oriented reforms will not be reversed. Cavallo expects that 6 percent per annum growth will become the norm for Argentina, and that it is well on its way to rejoining the industrialized world.

Foreign investors, while not as sanguine, began taking a close look at Argentina in the early 1990s. The country has long been a heavy exporter of capital, but confidence is returning now that Menem has demonstrated a firm commitment to economic reform. Argentina is now in the unusual position of attracting long-term capital inflows. A rash of takeover announcements by foreign companies has developed, in particular in the promising food and financial services sectors. In an especially welcome vote of confidence, General Motors has announced plans to return to Argentina, which it departed in 1978.

But while applauding the spectacular results achieved by Menem and Cavallo, it must be said that Argentina still has many obstacles to overcome on the runway to takeoff. First and foremost, domestic savings and investment levels are still much too low; the country needs to attract large-scale investment from capitalists still wary of its history of political, economic, and financial instability. The government will need to demonstrate that its newfound economic progress is based on economic fundamentals and not on a fixed exchange rate that has made Buenos Aires one of the world's most expensive cities. Also, the long-cosseted private sector may not be up to its new role as the engine of growth, leaving food production as the only sector in which Argentina enjoys an undisputed competitive advantage.

Pessimists fret that the spectacular growth of the past three years will prove to be ephemeral, founded on a consumer boom financed essentially by privatization revenues rather than on a sustainable base. Fundamentally, we can identify at least seven areas of concern as to the future course of Argentine reform: the need to deepen democracy; labor union challenges to the reform process; military opposition to the government; entrenched and high-level corruption; the difficulty of further fiscal reform; external debt; and exchange rate policy.[10] The most immediate concern is the sustainability of the hard exchange rate stance, especially in the wake of the Mexican peso collapse of early 1995. The peso's parity with the U.S. dollar has been the key element in the successful battle against inflation. However, it has also contributed to a soaring current account deficit, which is financed by highly volatile inflows of foreign capital. The IMF has warned Argentina to curb these capital inflows, which it fears will create excessive consumer demand and ultimately undermine reforms, but the government refuses. In the meantime, the fight against inflation is not over yet; while superlative by Argentine standards, underlying inflation is still well above the levels expected of an OECD country.

Finally, the base of reform may still be fragile. Vast social problems remain, while corruption and government scandals have been problematic. Consolidation of past gains will almost certainly imply an increase in social tensions and mounting pressure on Menem to ease his austerity policies. The labor sector is still an Achilles heel, with labor laws that make dismissals difficult and costly, and unions that wield power virtually throughout the economy. Menem also faces intense pressure to raise spending in order to fix the visibly collapsing infrastructure, an essential ingredient in attracting foreign investors. Moreover, Menem still faces strong and determined opponents to his reform plans, from labor and business groups to pensioners and teachers.

On the whole, though, Cavallo's optimism is not misplaced. Unlike previous plans, this time Menem and Cavallo have made serious efforts to cut the root causes of inflation—the government's budget deficit and unrestrained state spending—through deregulation, privatization, and fiscal discipline. Menem appears committed to spending the last part of his term consolidating these reforms, and looks increasingly likely to serve another term in 1995–1999 following changes in the Argentine constitution. Argentina still has a long way to go (by 1992 the economy was about the same size as in 1980, while the population had grown by 14 percent), but reforms and liberalization have set it on the right path.

Sequencing: Lucky, Again

Like other Latin American leaders, Menem is forced to maneuver in a tricky and idiosyncratic young democracy. But unlike Brazil, Argentina confers upon its president enormous strengths that help him to overcome the system. The constitution provides for a republican government, with separation of powers into executive, legislative, and judicial branches. It grants considerable power to the executive, however, and Menem's style of governing has exaggerated this process. Thus, the transition to democracy has not destroyed the executive's power to implement economic reform. Indeed, Menem himself has commented that economic "reforms would not have been possible without the presidential system that we have."[11]

Menem has spearheaded a battle to break down Argentina's traditional institutions without creating any new power sources to replace them. He is breaking up the unions, once the backbone of Peronist power, and has allowed the party to shrink in importance. At the same time, its Falklands defeat plus its disastrous handling of the economy while in office has led the military to accept a secondary role in national politics, while the business community broadly supports Menem's policies. Menem has emasculated the Supreme Court and judiciary, ignores Congress as much as possible, and rules by decree when necessary.

The weakening of institutions meant to check the power of the executive has created fears about Menem's authoritarian style, and the consolidation of political power in his hands. Writing in *The Wall Street Journal,* an Argentine attorney warns of "executive intrusion into the judiciary," pointing to occasions when the Supreme Court, lower federal courts, and the office of the public prosecutor have "been deprived of independent functioning."[12] Manuel Angel Broda, an economist, commented: "We have an enormous concentration of power now in Argentina. It is not in the Cabinet, or in the Economy Ministry or in Congress. It is directly in Carlos Menem. He has the kind of power that Charles de Gaulle had in France."[13] After ten years of democracy the congress and judiciary are undoubtedly weak, and are dominated by the all-powerful president. The opposition is divided and seeking about for a platform; the unions are weakened by privatizations and labor reform laws; the military is quiescent after its Falklands defeat and repeated budget cuts. While his party has effective control of both houses of congress, Menem's style of ruling by decree often removes legislation from the political realm in any case.

While critics may decry this trend on democratic grounds, in terms of economic results it contrasts sharply with the sorry experience of Brazil, where an emasculated president is at the mercy of the congress and judiciary. A striking example of this is the sweeping executive decree issued by Menem in late 1991, which singlehandedly does away with decades of regulations and controls that had crippled Argentine businesses. The executive decree circumvents Congress, which was not acting quickly enough to enact Menem's proposals. Local analysts worried that this was "the most aggressive demonstration of Mr. Menem's political power and an attempt by him to show that he will bypass Congress to bring the country out of years of hyperinflation, instability, and economic and political isolation, and make Argentina what Mr. Menem has often described as a 'first-world country.'"[14] It could be argued that President Menem is treading down the path of soft authoritarianism.

This is the path that no Brazilian president can take under its current constitution, and that world opinion was reluctant to allow Fujimori to take in Peru. Moreover, it is a sharp contrast to old-line Argentine politics. In the pre-Menem days, "the government's attempt to restructure industry became buried under the weight of a dense domestic network of powerful interests, which the government was totally unable to control." As in Brazil, the president became a "prisoner of powerful pressures."[15] It is Menem's good fortune—coupled with his skill in seizing the initiative and capitalizing on the public will for change—that he governs at a time when he is granted immense powers if he chooses to exercise them.

It is still unnerving, though, that the democratic context in which economic change is occurring in Argentina differs so dramatically from the experiences of Chile and Mexico. The sequence of reform does differ sharply from that of Latin America's two premier success stories, with Argentina having advanced first with political reform, which is followed by economic reform. The question of whether this reform will be successful remains a crucial one.[16] Recent erosion of the president's power vis-à-vis Congress is especially alarming in this regard. Congress appears to be less willing to rubber stamp Menem's initiatives, and the party too is less quiescent. As the aura of economic emergency recedes, so will Menem's power. However, the most critical economic reforms appear irreversible, and Menem is still an extraordinarily powerful executive. His re-election bid in 1995 may alarm die-hard democrats, but will strengthen the case for sustainable, irreversible market reform. Thus, Argentina will be able to embrace both democracy and economic reform simultaneously, unlike Brazil and Peru, although its "democracy" is much more authoritarian and dictatorial than pure democrats would wish.

Enough Food to Feed the World

Finally, Argentina is blessed with a good mix of natural resources for the 1990s. Although possessing abundant mineral wealth, including some oil, the country never had enough to make it heavily dependent on commodities exports like Venezuela and Peru. In a sense, it has the ideal level of oil—just about enough to make it energy self-sufficient, while not becoming a major oil exporter and thus being lured into that trap. Oil production is sufficient to meet domestic needs and produce a small exportable surplus. Reserves are estimated at seven to eight years of consumption, while gas reserves are equivalent to thirty years consumption at current rates and hydroelectric potential is enormous. Indeed, the elimination of state monopolies, removal of the regulatory burden, and liberalization of foreign investment laws have made energy one of Argentina's fastest-growing sectors. While some doubts have been raised about Argentina's ability to remain energy self-sufficient into the twenty-first century, this should not prove a serious constraint as improving technology can reduce energy input requirements.

The country's greatest strength, though, is in its agricultural sector. Endowed with abundant quantities of fertile land, Argentina boasts one of the world's lowest-cost farming sectors. It is more than able to feed itself; indeed, Argentina exports vast quantities of meats, grains, and oilseed to the rest of the world. Moreover, relatively low population growth (just 1.3 percent a year) will be a major advantage in an era of growing populations and hunger in the developing world. This ability to more than feed its domestic population, plus limited demand for energy imports, sets Argentina apart when compared to most other countries in the region.

All the Right Stuff

Of course, Argentina has had enough oil and food for years, but this did not help it to escape the abyss of economic disaster into which it tumbled during the 1980s. In some ways, food self-sufficiency may have prolonged the crisis: It encouraged Argentines to believe they could be more intransigent in negotiating with foreign creditors since they did not need new credits to import food from abroad. Gross mismanagement and corruption, combined with ideological leanings toward nationalism, statism, protectionism, and xenophobia, helped to obscure the natural strengths of the Argentine economy for many years, even decades.

Now, however, these strengths have the opportunity to assert themselves once more. Thanks to the comprehensive and deep economic reforms instituted by Menem and Cavallo, Argentina has a chance to compete in world markets once more. Blessed with abundant food and sufficient energy; supported by an educated and sophisticated urban workforce; and piloted by a strong executive—unshackled by normal democratic processes—Argentina is poised for takeoff.

Notes

1. For two excellent accounts of Argentina's developmental experience, see Paul Lewis, *The Crisis in Argentine Capitalism* (Chapel Hill: University of North Carolina, 1992), and William C. Smith, *Authoritarianism and the Crisis of the Argentine Political Economy* (Stanford, Calif.: Stanford University Press, 1989).
2. Peter G. Snow, "Argentina: Development and Decay," in Jan Knippers Black, ed., *Latin America, Its Problems and Its Promise* (Boulder, Colo.: Westview Press, 1991, 2nd edition), 510.
3. Scott B. MacDonald, Georges A. Fauriol, and Raul Luke, *Fast Forward: Latin America on the Edge of the 21st Century* (New Brunswick, N.J.: Transaction Publishers, forthcoming).
4. International Monetary Fund, *International Financial Statistics, August 1994* (Washington, D.C.: IMF, August 1994), 84.
5. Political Risk Services, *Country Reports: Argentina* (Syracuse, N.Y.: 1994).
6. See Andrew Dragoumis and Juan P. Villanueva, "EDESUR, Empresa Distribuidora Sur Sociedad Anonima, Argentine Company Report," *CS First Boston Fixed Income Research Report* (5 August 1994).
7. CS First Boston, *Prospects 1994: Q3* (July 1994): 55.
8. Snow, "Argentina," in Black, ed., *Latin America,* 510–12.
9. Jerry Haas, "The Argentine Miracle Continues," *Latin Finance* (June 1994): 82.
10. MacDonald, Fauriol, and Luke, *Fast Forward.* Edward Schumacher, "Shades of the Past in Argentina's Economic Boom," *The Wall Street Journal* (25 February 1994).
11. John Berham, "Menem Gets His Way on Route to Second Term," *The Financial Times* (14 December 1993).
12. Laura G. Koch, "Old Democratic Pitfalls Dog the New Argentina," *The Wall Street Journal* (27 March 1992).
13. Nathaniel C. Nash, "Argentina's President Turns Economics Into Poll Approval," *The New York Times* (16 February 1992).
14. "Argentina Deregulates its Economy," *The New York Times* (2 November 1991).
15. Monica Peralta-Ramos, *The Political Economy of Argentina* (Boulder, Colo.: Westview Press, 1992), 7.
16. MacDonald, Fauriol, and Luke, *Fast Forward.*

References

"Argentina Deregulates its Economy." *The New York Times* (2 November 1991).

Berham, John. "Menem Gets His Way on Route to Second Term." *The Financial Times* (14 December 1993).

Black, Jan Knippers, ed. *Latin America, Its Problems and its Promise., 2nd edition.* Boulder, Colo.: Westview Press, 1991.

CS First Boston. *Prospects 1994: Q3* (July 1994).

Dragoumis, Andrew and Juan P. Villanueva. *EDESUR, Empresa Distribuidora Sur Sociedad Anonima, Argentine Company Report.* CS First Boston Fixed Income Research Report (5 August 1994).

Fidler, Stephen. "Argentina (Survey)." *Financial Times* (17 May 1993).

Haas, Jerry. "The Argentine Miracle Continues." *Latin Finance* (June 1994).

International Monetary Fund. *International Financial Statistics, August 1994.* Washington, D.C.: IMF, August 1994.

Koch, Laura G. "Old Democratic Pitfalls Dog the New Argentina." *The Wall Street Journal* (27 March 1992).

Lewis, Paul. *The Crisis in Argentine Capitalism.* Chapel Hill: University of North Carolina, 1992.

MacDonald, Scott B., Georges A. Fauriol, and Paul Luke. *Fast Forward: Latin America on the Edge of the 21st Century.* New Brunswick, N.J.: Transaction Publishers, 1995.

Nash, Nathaniel C. "Argentina's President Turns Economics Into Poll Approval." *The New York Times* (16 February 1992).

Peralta-Ramos, Monica. *The Political Economy of Argentina.* Boulder, Colo.: Westview Press, 1992.

Schumacher, Edward. "Shades of the Past in Argentina's Economic Boom." *The Wall Street Journal* (25 February 1994).

Smith, William C. *Authoritarianism and the Crisis of the Argentine Political Economy.* Stanford, Calif.: Stanford University Press, 1989.

Snow, Peter G. "Argentina: Development and Decay." In Jan Knippers Black, ed., *Latin America, Its Problems and Its Promise. 2nd edition.* Boulder, Colo.: Westview Press, 1991.

4

Morocco: An African Tiger

Morocco usually evokes colorful images—the sun-bleached rooftops of Tangier, the desolate, yet impressive Riff mountains, and the sands of the Sahara. Thousands of European and American tourists flock to the North African country's beaches, while others jaunt into the interior cities of Fez and Marracach to examine the country's rich historical past. Beyond the tourist facade of juxtaposed Arab and Berber cultures, Morocco remains largely unknown. In the 1990s this North African country of 27 million people with a per capita income of $1,030 is on the move and plugging into the global economy.

This is not the way it has always been. Morocco in the early 1980s was regarded as a troubled debtor. The government's "Moroccanization" program, the adoption of an import substitution strategy, a heavy dependence on phosphate exports, and a growing reliance on external financing, led to an economic crisis in 1983. In that year, the real GDP growth rate slipped to 2.1 percent (versus a 3.2 percent annual population growth rate), and government-imposed austerity measures added to an unemployment rate that hovered around 30 percent. According to the World Bank, some 45 percent of the total population then lived below the poverty threshold of $238 per capita. Moreover, Morocco's deepening economic crisis had already provoked the violent 1981 demonstrations over increases in food prices. As if adding insult to injury, Morocco was forced to reschedule its commercial bank and official debt and begin the difficult process of economic restructuring.

By the 1990s the Moroccan economy has become a different animal. Prospects for strong economic growth have improved, external debt payments are no longer onerous, and the government's finances are largely under control. Direct foreign investment is also on the upswing—from $1 million in 1986 to $493 million in 1993.[1] Instead of being regarded as an economic disaster, Western investment banks were cautiously touting the advantages of Morocco over other developing countries.[2] The growing confidence about Morocco's prospects was further buoyed by the return of Morocco, after an absence of ten years, to international capital markets in July 1993 via a modest $25 million Euro-loan for Omnium North Africain. As of November 1994, Morocco's foreign exchange reserves stood at $4.2 billion, compared to $407 million in 1987.[3]

Despite concerns about official corruption, monarchial succession, and widening political participation, Morocco in the last decade of the twentieth century offers an impressive track record of stability, economic liberalization, and prudent management of external debt. Economic growth in the 1980–1991 period was a comfortable 4.2 percent, a pace well ahead of the population growth rate (2.6 percent for the same period). According to the World Bank, this puts Morocco's growth for that period ahead of such countries as Argentina (–0.4 percent), Brazil (0.6 percent), and neighboring Algeria (3.0 percent).[4] Despite an economic slowdown in 1992 and 1993 strong growth of 10 percent occurred in 1994. Moreover, unlike its North African neighbors Algeria and Egypt, Morocco has avoided the disruptive rise of radical Islamic forces. Like many of the first-generation tigers, Morocco is becoming attractive to foreign investment, has opted for an export-oriented growth strategy, and gradually is plugging into technology, largely driven by its close economic ties to Europe and the United States. Morocco in the 1990s has the potential to become a new tiger. Though problems exist that could preclude that advancement, Morocco is most likely to gradually progress through the decade to become a new tiger.

The Past as it Shapes the Future

Morocco sits on the northernmost shoulder of the African continent. Historically, it has been a crossroads of conquering armies and traders, ranging from the Carthagians and Romans to the Islamic forces that had burst from the deserts of the Arabian peninsula.[5] Enjoying periods of independence, what eventually emerged as Morocco was a regional power, holding sway over large parts of the African lands to the south. Morocco was also Islam's launching point for the conquest of the Iberian peninsula. Morocco, however, in the late nineteenth century underwent a profound decline and by the early twentieth century it was under the sway of European powers, especially France and, to a lesser extent, Spain. It was only in 1956 that Morocco was to regain its independence from the French.

During the struggle for liberation from French rule, the King, or Sultan, Mohammed V, played a critical role, which allowed him to make the monarchy *the* key institution in the new Morocco, over the claims of the nationalist movement, Istiqlal. The King was an able politician, consolidating his dominance over the political system. Mohammed V's position as national leader was also strengthened by his claims to be a direct descendant of the Prophet Mohammed—an influential Koranic title in a predominantly Muslim country.

When King Mohammed V died during a medical operation on 26 February 1961, he was replaced by Crown Prince Hassan II, then commander-in-chief of the armed forces. On 3 March 1961, when he formally succeeded his father, Hassan II's prospects for survival were hardly guaranteed. A wide range of forces, such as the political parties and a hostile Algeria, soon challenged his claim. The young King's intention to create a limited democratic system under royal tutelage, through a constitutional monarchy and a limited multiparty system, opened the door to challenges to royal authority.[6] After riots rocked Rabat, Fez, and Casablanca in 1965, the King was forced to close parliament. From that point forward, Hassan II proved to be a strong-willed authoritarian with a pragmatic side that allowed him some degree of flexibility in dealing with opposing forces.

While Hassan II was ruthless in stamping out opposition figures, such as Ben Barki, who it is alleged he had assassinated in Paris (creating a major international incident), the King allowed a degree of limited political freedom. This was achieved by maintaining his father's policy of making the monarchy central to the nation's development. The challenges to Hassan II included a number of riots, as in 1965, 1982, 1984, and 1990 and two military coup attempts, in 1970 and 1972, as well as leftist plots in 1973. The military coup attempts were severe challenges, the first resulting in the execution of a number of high-ranking officers, while the second involved the regime's right-hand man to the King, Oufkir. In each case, the King always emerged victorious, his power base enhanced in a delicate balancing game.

Within the context of the monarchy's political dominance, economic policy evolved largely at the dictates of the King and his advisors. It is important to underscore that the starting point of economic development in Morocco and policies implemented by the government were conditioned by three elements shared by a majority of Middle Eastern countries. Richards and Waterbury noted it was strongly felt by new national leaderships that the European imperialist legacy was backwardness, which had three components: "(1) the agrarian trap, whereby the economy is mired in the production of cheap agricultural commodities requiring an unskilled work force; (2) the system of production is perpetuated by denying education and the acquisition of modern skills to all but a privileged few; and (3) the trap is sprung through the international division of labor into which the backward agrarian economies are forcibly integrated."[7] Because of these conditions, much of the Middle East opted to use the state to mobilize resources, provide training, and create new societies that were to be better educated, more urban, and at a higher level of industrialization. An additional factor of the preferred use of the state has been the lack of legitimacy and capacity of the private sector to be the primary locomotive in the economy. Morocco at independence was largely characterized by such conditions.

Under Mohammed V and then Hassan II, economic policy was guided by import substitution and an interventionist state. The economy was founded on agricultural performance and the export of phosphates and phosphate byproducts, stimulated by public sector expansion and the need to maintain growth above the pace of the population, which averaged 2.5 percent from 1965–1980. The concern over population growth has been an ongoing factor in Morocco's development, especially as the heavy weight of new entrants into the job pool has put pressure on the state to provide education and training. At the same time, Morocco's abundant labor kept wages low and gave its exports a comparative advantage in international markets. Morocco also benefitted from the remittances of its sizeable emigrant communities in Europe.

Morocco's economic development began a roller coaster ride in 1974–1975 when phosphate prices rose precipitously due to the overall spike in commodity prices. While Morocco enjoyed economic expansion in the 1960s, the government had pursued conservative fiscal policies; in the 1970s the large and sudden infusion of export revenues provided the government the false sense of security to engage in a more free-spending track. Exports, however, did not provide all the capital required for the King's government plans—some costly and grandiose projects. The solution was to borrow from commercial banks in North America, Europe, and Japan. To both commercial bankers and their Moroccan borrower the future source of repayment, phosphate earnings, appeared secure. The first Organization of Petroleum Exporting Countries (OPEC) oil shock helped establish a perception that natural resources, such as oil and phosphates, would only increase in value. The miscalculation made by many analysts was to de-emphasize the ability of many industrialized countries to reduce imported energy and to find alternatives to higher-priced commodities, such as phosphates and oil.

Part of Morocco's spending program included weaponry. The desire to overhaul the economy, coupled by an arms buildup, stimulated the expansion of the country's external debt. The arms buildup was caused by the Saharan war with the Polisario Liberation Front (PLF), which commenced after Morocco's annexation of former Spanish Sahara in 1975. The PLF were not overcome and launched a guerrilla war against Morocco and Mauritania (which had participated in the annexation taking a southern third of the territory). The PLF managed to knock Mauritania out of the struggle, but Moroccan forces assumed control of that part of the former Spanish Sahara. The use of southern Algeria as a major base of Polisario operations did little to reduce tension between Algiers and Rabat. By 1979, Morocco's external debt had grown from $953 million in 1970 or 24 percent of the country's GNP to $9.9 billion or 64 percent of GNP. Total debt service swelled as well, reaching 23 percent of goods and services exports.

By the late 1970s the Moroccan economy was in trouble. Despite the state's expanded role in the economy and pouring of money into such diverse sectors as construction, sugar, fertilizer plants, and petroleum refining, there was a failure in increasing exports to offset the rise in imports. The external accounts began to shift against Morocco, with interest payments contributing to large current account deficits of over $1 billion annually from 1976–1979. Austerity measures implemented in the 1978–1980 period failed to halt the economic slide. High oil prices further weakened Morocco's economic position, which was reflected by the continuation of current account deficits of over $1 billion in 1980 and 1981. Food riots, sparked by higher prices, rocked the country in 1981. Although Morocco staggered through 1982, 1983 was a year of reckoning. Forced to reschedule its external debt, Morocco's leadership decided to change direction on the economic front.

Beginning in 1983, Morocco embarked upon economic programs aiming to turn around the deficits in the external accounts, bring state spending under control, provide an export orientation for the economy, and stimulate the private sector. Additionally, efforts were to be made to reduce the dependence on phosphates and agriculture. All of this, of course, helped provide better management of the nation's economic resources while helping to bring the external debt under control. To achieve these ends in the 1980s Morocco adopted a more flexible exchange rate policy, opened up its markets to foreign goods (including membership in the GATT, the General Agreement on Tariffs and Trade), lowered customs duties, and reduced bottlenecks, which had prevented the effective distribution of goods. Other steps included the elimination of licensing requirements or taxes for private exporters, the dismantlement of the state-run Office de Commercialisation et d'Exportation (which had a monopoly on farm produce and processed foodstuff exports), the implementation of stronger tax and export incentives, guarantees for repatriation of capital, and the removal of remittances of dividends.[8]

Another important reform regarded foreign ownership. In 1985 the Moroccan government abandoned the policy of making foreign investors accept Moroccan majority partners in joint ventures. This was followed in subsequent years by the removal of all restrictions on the degree of foreign ownership in an enterprise. The response of foreign investors to these changes was positive. In the 1982–1988 period foreign investment grew by an average of 7 percent annually, with the last year witnessing a 25 percent expansion, reaching $140 million.[9]

Although Morocco did not emerge as a new tiger by the end of the 1980s it did evolve into an understated economic success story. Throughout most of the mid-1980s to the early 1990s Morocco achieved relatively high growth and enjoyed low inflation (6.7 percent in the 1982–1991 period). Over the 1986–1991 period, Morocco's real GDP growth averaged 5.1 percent, exceeding Mexico (2.5 percent), Argentina (1 percent), Brazil (1.9 percent), and oil-exporting countries such as Algeria (0.9 percent), Venezuela (3.6 percent), and Nigeria (4.2 percent).[10] Additionally, Morocco's external accounts have become more manageable and the country's foreign exchange reserves were at $3 billion in 1992, a 25-fold increase over the past decade (see table 4.1).[11]

Morocco's economic program weathered the 1990–1991 Middle East crisis caused by Iraq's ill-fated August 1990 invasion of Kuwait. Public

TABLE 4.1
Morocco Basic Economic Data

	1988	1989	1990	1991	1992	1993	1994E
Real Growth	10.4	2.5	3.7	5.2	-4.4	-1.1	10.0
CPI (%)	2.4	3.1	6.9	8.0	4.9	5.9	4.5
Exports ($mn)	3,608	3,312	4,210	4,277	3,956	3,850	4,010
Imports ($mn)	4,360	4,991	6,282	6,253	6,692	6,820	7,040
Current Account Bal. ($mn)	467	-790	-200	-396	-427	-690	-490
Int'l Res.	547	488	2,066	3,100	3,504	3,410	3,610
Total External Debt ($ bn)	21.1	22.1	22.5	22.9	23.2	22.5	23.0
Debt Service Ratio (%)	26.1	33.1	21.9	27.6	26.6	33	34

Source: Economist Intelligence Unit, World Bank, IMF.

sentiment in favor of Saddam Hussein's "standing up to the West," and the sending of humanitarian aid to Iraq by members of the royal family were balanced by the King's detachment of 1,000 troops to help the Gulf Arabs. However, a temporary increase in energy costs and the adverse effects of drought on the economy left a pool of dissatisfied people, who were stirred by the Gulf War. The December 1990 disturbances in Fez and a few other cities signalled to the King's government that a degree of social tensions existed due in part to difficult economic conditions.[12]

Aware of the need to deal with social issues, such as unemployment and job training, the King launched a new program in January, 1991. In a well-publicized speech, Hassan II announced increases in the statutory minimum wage levels, salaries for low-paid government workers, and higher family and social welfare benefits. The program also included an ambitious job creation program aimed at employing 50,000 jobless university degree and diploma holders by July and creating a total of 100,000 new positions by year-end 1991.

The decision to boost public spending came despite the government's earlier desire to maintain a low budget deficit. In 1989, the budget deficit was 5.7 percent of GDP, regarded as too high by the IMF; in 1990 it was reduced to a little over 3 percent.[13] For 1991, it remained around 3 percent, but creating a social cushion was regarded as politically important, especially considering the problems related to the economy in neighboring Algeria. Despite tough measures in 1992, 1993 witnessed less restrictive fiscal and monetary policies. The situation improved again in 1994, falling under 3 percent of GDP in 1994.

Morocco also benefitted from the growth of the informal sector of the economy, especially in the textile industry. Although the wonders of the informal sector should not be overstated and accurate data is largely nonexistent, it is probable that this part of the economy helped absorb some of the country's unemployed. The rapid expansion of the informal sector in the late 1980s and early 1990s was partially fuelled by the low costs of production and by workplace regulations that constrained industry in the formal economy. One analyst suggested that the combination of new foreign and local private sector investment in the economy helped create 45,000 jobs in the private sector in 1990 alone.[14] The combination of job creation in the official private sector and expansion in the informal sector helped offset the effects of the state's low profile in stimulating economic activity and employment as well as providing a partial buffer to potential social discontent. However, in 1991 the situation clearly called for a more active state role in muting public unrest related to problems at home with unemployment and other ill effects of structural adjustment and foreign events, including the Gulf War and the rise of Islamic fundamentalism in neighboring Algeria.

Morocco emerged from the Middle East crisis in a stronger position than many other countries, especially those that openly sided with Iraq, like Jordan. A key factor was the appreciation for Morocco's assistance from Saudi Arabia: in 1991 the Gulf state forgave $3.6 billion of official debt owed by Morocco (equal to almost 90 percent of Morocco's $4.1 billion commercial bank debt as of 1992).[15] Saudi appreciation was also demonstrated by free oil for six months and $590 million in foreign aid to compensate for the effects of the Gulf crisis. Further Gulf Arab largesse came from the Kuwait Fund for Arab Economic Development and the Arab Fund for Economic and Social Development, which earmarked project finance for the improvement of water systems and roads.

Nor was Morocco forgotten by the West in the aftermath of the Gulf War. Already receiving a substantial amount of official assistance from France, Morocco got a commitment from Spain of $1.3 billion over a multiyear period. Morocco also receives a small amount of assistance from the United States in the form of project financing and military credits. For Europe and the United States, Morocco's position as a moderate government in the Middle East and North Africa has been an important determining factor in maintaining economic assistance.

Tiger Factors

Morocco's tiger factors include the sequencing of economic reforms (with a new round looming—banking and financial sector reforms, full dirham convertibility, and privatization), the diversification of the economy, strong leadership, and a gradual improvement in the country's technological foundation. It is these factors that will determine Morocco's economic future.

Strong Leadership

King Hassan II has provided strong leadership for Morocco and is a key factor in its gradual transformation from troubled debtor to potential tiger. As Alan Richards and John Waterbury noted of Hassan II: "He assiduously followed the divide-and-rule tactics of his father, weakening parties, trade unions, and regional interests, but never destroying them or pushing them out of the political arena...King Hassan puts himself forward as the final protector of all interests; the parties against military intervention; the military against civilian bungling; the Berbers against the Arabs; and the Jews against the Muslims. Hassan II may not be loved, but most Moroccans now do believe that the monarchy is vital to the country's stability."[16] Morocco's political system in the 1990s is soft-shelled authoritarianism. While elections are held, parliament conducts its debates and passes legislation, and the opposition is allowed to vocalize criticism of the government, there are well-defined limits. The King is clearly at the apex of a hierarchical political system, a position greatly strengthened by the Alawite royal family's sharifian status (the king is an iman or spiritual leader, i.e., the commander of the faithful).

Though a degree of bargaining typically occurs between the various political actors—the political parties, military, students, labor unions, and workers in the agricultural sector—the King usually has the final say. There are coercive mechanisms to ensure the King remains above the fray and his policies are carried out. I. William Zartman noted: "Morocco is a polity where an overly active opponent will be offered the choice of jail or a high position if he is important enough, but an ordinary recalcitrant will merely get beaten up. State violence and control are an intimate part of the system, and they operate alongside other elements of systemic restructuring such as agenda control, cooption, and orchestration, not as an alternative."[17]

Zartman's observation was reflected by the Hassan regime's willingness to wipe out a generation of young military officers with the Kenrita trials in 1972 (following an abortive coup), as well as its response to radical student demonstration in 1973. The repressive side of the monarchical government also revealed a disregard of due process and judicial niceties with small numbers of political prisoners disappearing for long periods without a trial.

The King's power is based on royal institutions, the military, and allied political parties, as well as his religious position as iman. This strong leadership has provided a high degree of political stability in sharp contrast to most of Africa. The high degree of political stability in turn has provided continuity in economic policy, allowing the government to pursue a gradual structural adjustment process instead of shock treatment. The King's ability to orchestrate Morocco's political life has also resulted in a policy implementation environment where opposition to economic change remained weak and ineffectual, allowing the national leadership to foster a consensus within most of the elite and propagandize the positive elements of the reforms to the population. Most importantly, this has allowed the government to create a sizeable constituency that is supportive of the reform process. All of this was and is based on a quasi-authoritarian foundation.

The trick for the King's government will be to gradually transform the country's political system into one that can better accommodate the growing range of economic freedoms. The June and September 1993 parliamentary elections were a critical step in what may evolve into a gradual political liberalization. Although the King was forced to appoint a government dominated by technocrats due to the opposition parties' refusal to participate in the ruling coalition, the postelection government is more broad based, reflecting a desire on the part of the King to spread the responsibility of government further and push to maintain a national consensus on developmental objectives. In this, the King's government will have to work more closely with the newly elected Chamber of Representatives. That body's role is currently advisory and dominated by center-right parties, but can be expected to play a more important role in the country's future direction. The King has also indicated that he would prefer to hand over power at some stage to the crown prince, but remain the power behind the scenes, hence providing a degree of stability in the transfer of authority. This may be central to maintaining political stability in Morocco as the ability of the crown prince to govern, in tandem with the parliament, may prove to be a difficult task.

Morocco's recent political reforms indicate that the Royal family has not missed the events that shaped much of the Middle East and North Africa, in particular the demand of populations for some form of greater accountability from their governments. The failure of Algeria's socialist authoritarianism and the resentment many Arabs possess for the Gulf monarchies was not lost on the Moroccan regime. Algeria's troubles have been related to the rise in popularity of the Muslim party, Front Islamique Salut (FIS). The FIS gained considerable popularity not because of the attractiveness of its message of returning Algeria to Koranic law, but due to its focus for opposition to a corrupt and ineffectual regime devoid of any new ideas of governing. Moreover, Algeria's leadership proved weak in the face of new challenges in the post-oil age in the 1980s and the consensus of the national elite became a facade. For Morocco to avoid Algeria's troubles, a gradual political opening is required, but a political opening in which the major actors largely agree upon the direction for the nation. In the 1990s Morocco appears to have a degree of national consensus within the political elite that national development is evolving along the right track.

Continued Sequencing of Economic Reforms

Morocco has accomplished much in sequencing its economic reforms. The process of abandoning the import substitution strategies in the 1980s and the transformation of the economy have come in stages and greatly reduced the burden of the country's external debt (at $22.5 billion in 1992, most of which was owed to official debtors, not commercial banks). Although the pace has not always met expectations (as with privatization), the sometimes gradual pace of reform has produced considerable results without the disrupting effect of shock reform programs pursued in much of Central and Eastern Europe in the late 1980s and early 1990s. The path ahead will be one of continued sequencing of reforms, albeit at a faster pace.

One of the government's long-standing goals was achieved in January, 1993 when dirham convertibility was introduced for current international transactions.[18] Although the government retains certain authority over exchange controls, foreign exchange licenses are no longer required and commercial banks can now supply the funds directly. The remaining prohibitions include the holding of offshore accounts and limits remain on Moroccan residents for foreign exchange allocations for business and tourist travel. (The value of Morocco's dirham is determined on a trade-weighted basis, reflecting a basket of currencies of the most important trading and tourism partners.)

The King outlined his country's future development goals in June, 1993. The four primary objectives were to create investment incentives and reduce bureaucratic delay of new investment; reform the financial system, in particular the banks and the Casablanca Stock Exchange (CSE); to quicken the pace of privatization; and finally, to establish proper and transparent business practices. These reforms are highly important due to stiff competition for foreign investment, in particular from Central and Eastern Europe as well as China, Malaysia, Thailand, and Vietnam.

The privatization program was originally approved in 1989, but it was not until October, 1992 that some tentative progress was made (four companies were sold). The program accelerated in 1993 with companies and hotels being sold, while fifteen companies and hotels were privatized in 1994 (as of November). Ultimately the program will sell 112 companies by 1995 with government hoping to raise between \$2.4–2.9 billion in revenues through sales.[19] The sell-off candidates include fifteen hotels, petroleum distributors, chemical companies, contractors, brewers, and transport firms. If the privatization program is successful it will not only raise a substantial amount of revenue for the government, it will also help stimulate activity in the CSE while avoiding an upswing in unemployment—a factor common in the privatization process in Central and Eastern Europe and parts of Latin America.

Financial reform represents a tough challenge for Morocco. The CSE as of 31 March 1993 had a market capitalization of \$2.7 billion. However, the CSE is open only one hour a day and the number of brokers is below twenty (dominated by banks). Of the sixty-nine listed companies only five are actively traded. To complement the privatization program and breathe life into the CSE, four reforms will be implemented: mutual fund vehicles will be introduced; the CSE will become a private sector organization and run by brokers instead of by the state; requirements will be put in place for enhanced public information disclosure, part of which will be accomplished by the establishment of a Securities Exchange Commission; and lastly, from 1992 onward, companies listed on the CSE will be audited. Data on Moroccan companies in the past has been spotty and investors had to be wary. This set of reforms would provide a more structured business environment along accepted international standards and reduce chances of scandals caused by financial fraud and shoddy accounting practices.

Morocco's banking sector is also in the process of reform. In 1993, the government adopted the Basle Committee's capital adequacy standard of 8 percent capital reserves for loans outstanding. Although this move caught some banks in difficulty, the adoption of such standards, originally limited to the major industrialized countries, puts Morocco's banking system in a stronger and safer position.[20]

The Tech Factor

Morocco's implementation of technology has been gradual and is still at a slow pace. Despite this cautious approach to introducing new technology, Morocco has made certain important advances. A significant breakthrough came in 1982 in agriculture. In that year, the Moroccan government started to promote farm mechanization by removing all duties and taxes on farm equipment. The government also provided credit for such purchases. The result was, as Richard and Waterbury noted: "The real cost of a tractor fell by 30%, and not surprisingly, the number of tractors rose from 23,000 in 1978 to 40,000 in 1986. These policies, combined with growing incomes, accelerating demand for animal products, and, in some cases, shortages of young adult male labor have contributed to the rapid tractorization of the region."[21] This introduction of better technology has clearly helped make Morocco a significant exporter of agricultural products to the European market.

Although there has been no substantial increase in overall manufacturing relative to GDP since 1980, the industrial sector has not been idle in terms of adapting new technology. Morocco's manufacturers are aware of the need to move up the technology ladder, in large part because the country's international competitors are seeking to do exactly the same. New management expertise has been added to the workplace, intermediate training programs have been introduced, and strategic alliances with foreign firms are being sought. Improvements are already being felt. As Francis Ghilès noted in July, 1993: "There are encouraging signs. In the last five years, the export of value-added products has doubled, moving from 22 to 30 percent of the total."[22] Further improvement in this sector is expected as new technology is being introduced.

The Moroccan government is aware of the need to upgrade its tech infrastructure. Consequently, among new capital projects launched in 1992 were a "substantial improvement" in telephone line availability to meet rising demand, thermal power plants to reduce the dependence on outside energy sources, and youth investment training programs to provide education for new industries.[23]

Improvement in communications is a high priority to the Moroccan government. The Office National des Postes et Télécommunications in its 1993–1997 plan will expand the percentage of automatic exchanges to 99.5 percent; a network of fiber optic cables is being extended throughout the country; a cellular mobile radio telephone system is being introduced; and communications with Europe, Morocco's major market, are being improved.[24] The total cost of the upgrade of the national telecommunications system is expected to be $1.2 billion, largely financed by the World Bank and other multilateral sources. The significance of this project was noted by the Economist Intelligence Unit: "Telecommunications development is a key element in Morocco's bid to transform itself from a rural-based developing country to a modern business-centered economy."[25]

The European Factor

Morocco's location south of Europe has brought a number of comparisons to the relationship between Mexico and the United States. As *The Economist* (9 January 1993) noted: "Yet in Rabat these days government officials clearly enjoy talking about the Mexican example. Like Mexico, they say, Morocco stands ready to throw open its protected markets to free trade from the rich north. And like the United States, the argument continues, the European Community [known as the European Union or EU after December 31, 1993] should see that its own interest lies in being generous towards its poor southern neighbor."[26]

Morocco is negotiating with the EU to establish by year-end 1993 or early 1994 a ten- to twelve-year transitional period to create a free trade zone. Already 60 percent of Morocco's exports are to the EU, while 55 percent of its imports come from the Community. It is probable that Morocco and the EU will come to an agreement that will strengthen the North African country's economic position.

Although European countries have concerns about the mobility of Moroccan labor (an estimated 700,000 Moroccans already work in the EU), concerns about the stability of the country on Europe's southern flank is great. This concerns stems from apprehension over illegal immigration and the expanding role of Morocco as both a transit point of hard drugs (cocaine and heroin) and marijuana and hashish to the EU. In the early 1990s, Morocco supplied Europe with 27 percent of all the cannabis consumed—the largest identifiable source.[27] The reasoning behind helping Morocco in its development is simple—if Europe does not assist the King's government in modernizing his country and providing its population with a better standard of living, the threat is that many Moroccans will opt to move to Spain, France, Belgium, or Germany in any fashion they can. Another concern for Europe is Morocco's geostrategic importance as a bulwark against the rise of Islamic radicalism and the political destabilization of the North African countries. Although a plethora of issues complicate the consummation of an agreement (including Morocco's spotty human rights record), European governments are likely to find the logic of what their local constituencies demand in terms of reduced immigration and concerns about crime, often perceived as related to newcomers.

Swing Factors

Political Sequencing

The sequencing of political reforms is perhaps the greatest challenge facing Morocco as its economy continues to gain momentum. Along these lines, Hassan II's governments have quietly embarked upon a political opening, which seeks to allow greater participation in the system. This should by no means be regarded as a departure into a complete constitutional monarchy along the lines of Spain or the United Kingdom, but a moderation of some of the regime's more autocratic dimensions. The government has closed Tazmamert, the country's notorious political prison; a new constitution was implemented in 1992; and a

Human Rights Commission was established. The new constitution contained language about human rights and was approved by a national referendum (though the approval rating was 99 percent!).

The King's efforts at presenting a more "democratic" Morocco were unfortunately undermined by the blatant rigging of municipal elections in October, 1992 and the postponing of parliamentary elections, originally scheduled for April, 1993. However, the June, 1993 elections were surprisingly open and with few cases of outright abuses. Moreover, the June contests were peaceful and more representative and credible than previous outings to the polls with only 63 percent of voter turnout.

In terms of political sequencing two options have been advanced for the pace and scope of change. One view is that the parliamentary elections (as well as the new constitution) were an attempt by the King to hand more power to the parliament to broaden the decision-making process and include opposition forces in the government.[28] This implies that the King will gradually devolve power in his hands to other responsible members of the national elite. Although the King's power would be reduced, he would still continue to orchestrate the country's political life and remain a unifying national figure.

The second view is that the King would prefer to relinquish power to his son, Crown Prince Sidi Mohammed, during his lifetime.[29] This would provide continuity in the national leadership, with perhaps a degree of greater authority going to the parliament. However, Hassan II would remain (as in scenario one) a powerful figure in the background. This situation has certain echoes in the Singaporean experience with Lee Kuan Yew officially resigning as prime minister, but maintaining a cabinet position and holding power behind the scenes for the eventual successor to national leadership of his son.

Whatever the inclinations of the King, Morocco will undergo a degree of political change in the 1990s and into the first decade of the next century. It will be critical that reform follow reform and that the elites and citizens both understand the new rules of the game. Morocco has been down the road of political liberalization before and halted because of the King's nervousness with the rapidity in which the political situation threatened to slip out of control—his control. Therefore, for political sequencing to succeed, all parties will have to demonstrate some restraint. Without that restraint, the hard line of the King's government could return, which in turn could cause political instability, which in turn could jeopardize the economic achievements.

Education and Poverty

Morocco faces a major challenge in the area of education; this is clearly a major swing factor in determining the country's pace of development. According to the World Bank, total adult illiteracy is 51 percent of the population.[30] Morocco's illiteracy rate is higher than a number of other developing countries: for example, Cameroon's 46 percent, Paraguay's 10 percent, and Tunisia's 35 percent. Morocco finds itself ahead of such economic "dynamos" as Nepal (74 percent), Chad (70 percent), and Sudan (73 percent), and on a par with India (52 percent), Nigeria (49 percent), and Egypt (52 percent). All of this means that Morocco is far behind the first generation of tigers, most of which have illiteracy rates below 5 percent. The educational picture is further complicated by greater illiteracy in the countryside and the higher rate of illiteracy among women (62 percent).

If Morocco is to make the leap to a higher stage of economic development it will have to improve its educational track record. Without coming to terms with inadequate educational facilities, upgrading the national circulum to match new international developments, and promoting new tech skills, Morocco risks being stuck as a pool of unskilled and semiskilled industrial labor and a supplier of commodities for the global economy.

The reductions in government spending in the 1980s have meant that the quality and availability of free education declined by the end of the decade. The Moroccan government in the 1990s has gradually come around to the point of view that more spending is necessary in the educational field if the country's progress is to continue. Accordingly, the Moroccan government and the World Bank channelled a large part of the proceeds of a structural adjustment loan of $275 million in 1992 toward education and health care for the poor.[31] In 1992 and 1993, therefore, increases were made in the allocation of spending on education.

Poverty is another major social problem facing Morocco. The causes are high unemployment, high population growth, and inadequate state facilities in education, health, and sanitation. Politically, this condition, especially in the slums referred to as *bidonvilles,* creates a pool of volatility that can be easily tapped. As with education, the government is seeking to address the issue: the 1993 budget provided a 10 percent increase in public spending.

While poverty is indeed a national problem, improvements in the standard of living have occurred in Morocco. The per capita income (a broad indicator) has doubled from $600 in 1985 to $1,100 in 1992, some of which has occurred at the lowest socioeconomic levels of Moroccan society. In 1991 only 13 percent of the population was living in absolute poverty.[32]

Drought

Morocco's economic growth has been partially dependent on the weather. This was reflected as recently as 1992–1993 when drought hurt the country's economic performance, in particular its agricultural sector. While the nonagricultural sector grew steadily in this period, the weight of drought caused a disappointing real GDP output. Morocco's sensitivity to drought stems from the fact that close to 90 percent of its agriculture is rain fed. Although drought in of itself will not derail the reform process in Morocco, it can contribute to the country's problems, in particular hurting its ability to feed itself and to gain export revenues. Clearly, Morocco must move more toward irrigation to buffer the ill effects of drought. Prospects for growth clearly have picked up in 1994 and early 1995 as drought has receded as a problem.

Conclusion

Morocco has achieved a certain momentum in the economic reform process that differentiates it from most African and Middle Eastern countries, such as Algeria and Nigeria. While problems exist with poverty, bureaucratic inefficiencies, and corruption, the scope is not endemic and the problematic side is controlled (sometimes harshly) by the government's security forces. More importantly, the economic reforms embarked upon in the 1980s have pushed Morocco to a new stage of development. This new stage may prove an even more difficult one, entailing political liberalization, monarchial succession, a new round of economic reforms, and further economic integration with the EC.

Short of revolution (Islamic or otherwise), the assassination of the King, or *force majeure*, Morocco's development as a new tiger is likely to continue through the 1990s.[33] Morocco is quietly achieving an economic success, which is beginning to be noticed by investors, especially when shopping throughout the rest of the world for a good investment environment.

Notes

1. International Monetary Fund, *International Financial Statistics, February 1995* (Washington, D.C.: International Monetary Fund, February 1995), 386.
2. See for example, Paul Luke and Morgan Grenfell, *Arbitrage Opportunities in Moroccan and Polish Debt* (16 March 1993) and, by the same bank, *Morocco: A Gradualist Approach to Adjustment* (19 May 1992). Also see Joyce Chang, et al., *Morocco: An Oasis of Investment Opportunity* (June 1992). Additionally, the Templeton Emerging Markets Fund in 1993 bought three Moroccan stocks—Omnium North Africa, Banque Commerciale du Maroc, and Wafabank—worth a few million dollars. See Ted Merz, "Templeton Emerging Markets Manager Nibbles at Moroccan Stocks," *Bloomberg News Wire* (16 November 1993).
3. International Monetary Fund, *International Financial Statistics* (February 1995): 385.
4. World Bank, *World Development Report 1993* (New York: Oxford University Press, 1993), 240–41.
5. For a history of Morocco see Bernard Lugan, *Histoire du Maroc: Des origines à nos jours* (Paris: Criterion, 1992).
6. George Joffé, "The Maghrib," in Peter Sluglett and Marion Farouk-Sluglett, *The Middle East: The Arab World and Its Neighbors* (London: The Times, 1991), 202. As the author noted of Morocco's early attempt at parliamentary government, "Parliamentary politicians exploited their positions for personal gain, and the extremist positions adopted by the political parties prevented parliament from carrying out its obligations under the constitution."
7. Richards and Waterbury, *A Political Economy of the Middle East,* 186.
8. John Roberts and Scott B. MacDonald, "North Africa and the Middle East," in Scott B. MacDonald, Margie Lindsay, and David L. Crum, ed., *The Global Debt Crisis* (London: Pinter Publishers, 1990), 79–80.
9. George Joffe, "Morocco: A Survey," *South* (September 1989): 89.
10. Chang et al., *Morocco: An Oasis of Investment Opportunity,* 5.
11. Ibid.
12. On 14 December 1990 a general strike was called in Fez by two of the Morocco's three largest trade unions. The strikes gave way to serious riots, which tapped considerable discontent with high unemployment, especially among the urban young. There was serious damage done in Fez, including the gutting of one of the city's three five-star hotels. In all there were two days of serious rioting that also occurred in Tangier and Agadir.
13. Sophie Bessis, "Le Maroc face à lui-même," *Jeune Afrique* (12 March 1991): 34.
14. Sophie Bessis, "Le Maroc face à lui-même," *Jeune Afrique* (March 1991): 32.
15. Chang, et. al., *Morocco,* 3.

16. Alan Richards and John Waterbury, *A Political Economy of the Middle East: State, Class, and Economic Development* (Boulder, Colo.: Westview Press, 1990), 319–20.
17. I. William Zartman, "King Hassan's New Morocco," in I. William Zartman, ed., *The Political Economy of Morocco* (New York: Praeger, 1987), 25.
18. Economist Intelligence Unit, *EIU Morocco Country Report* (Second Quarter 1993): 17.
19. Economist Intelligence Unit, *EIU Morocco Country Report 1* (1993): 14.
20. "Maroc-Banques: plus de fonds propres moins de risques," *Jeune Afrique* (24 December 1992–6 January 1993): 7.
21. Richards and Waterbury, *A Political Economy of the Middle East,* 175.
22. Francis Ghilès, "'Model Patient' Morocco Keeps Medicine Down," *Financial Times* (23 July 1993): 4.
23. Economist Intelligence Unity, *EIU Morocco Country Report 1* (1993), 13.
24. Ibid, 12.
25. Economist Intelligence Unit, *Morocco Country Report* (Second Quarter, 1993), 26.
26. "Morocco: M&M," *The Economist* (9 January 1993): 37.
27. François Soudon, "Guerre à la drogue!" *Jeune Afrique* (13 January 1993): 17.
28. Joyce Chang, et al., "Moroccan Parliamentary Elections: A Step Toward Political Opening," *Solomon Brothers Emerging Markets Research North Africa* (23 July 1993).
29. Advanced by Paul Luke, "Morocco: Research Update," *Morgan Grenfell* (17 February 1993): 3.
30. World Bank, *World Development Report 1992* (New York: Oxford University Press, 1992), 218.
31. Francis Ghilès, "Morocco still cautious despite economic progress," *Financial Times* (10 August 1992): 4.
32. Solomon Brothers, 11.
33. Francis Ghilès, "Revival of Interest in Moroccan Investment," *Financial Times* (16 July 1993): 19.

References

Bessis, Sophie. "Le Maroc face à lui-même." *Jeune Afrique* (12 March 1991).

Chang, Joyce, John F. H. Purcell, Valerie Chang, Dirk W. Damrau, and Ernest W. Brown. "Moroccan Parliamentary Elections: A Step Toward Political Opening." *Solomon Brothers Emerging Markets Research North Africa* (23 July 1993).

Chang, Joyce, John F. H. Purcell, Dirk W. Damrau, and Ernest W. Brown. *Morocco: An Oasis of Investment Opportunity* (June 1992).

Economist Intelligence Unit (London). *EIU Morocco Country Reports* (2nd quarter 1994).

Ghilès, Francis. "'Model Patient' Morocco Keeps Medicine Down." *Financial Times* (23 July 1993).

______. "Revival of Interest in Moroccan Investment." *Financial Times* (16 July 1993).

______. "Morocco still cautious despite economic progress." *Financial Times* (10 August 1992).

International Monetary Fund. *International Financial Statistics.* Washington, D.C.: International Monetary Fund, [monthly].

Joffe, George. "Morocco: A Survey." *South* (September 1989).
Lugan, Bernard. *Histoire du Maroc: Des origines à nos jours.* Paris: Criterion, 1992.
Luke, Paul. *Morgan Grenfell, Arbitrage Opportunities in Moroccan and Polish Debt* (16 March 1993).
______. "Morocco: Research Update." *Morgan Grenfell* (17 February 1993).
MacDonald, Scott B., Margie Lindsay, and David L. Crum, ed. *The Global Debt Crisis.* London: Pinter Publishers, 1990.
"Maroc-Banques: plus de fonds propres moins de risques." *Jeune Afrique* (24 December 1992–6 January 1993).
Morgan Grenfell. *Morocco: A Gradualist Approach to Adjustment* (19 May 1992).
"Morocco: M&M." *The Economist* (9 January 1993).
Richards, Alan and John Waterbury. *A Political Economy of the Middle East: State, Class, and Economic Development.* Boulder, Colo.: Westview Press, 1990.
Sluglett, Peter and Marion Farouk-Sluglett. *The Middle East: The Arab World and Its Neighbors.* London: The Times, 1991.
Soudon, François. "Guerre à la drogue!" *Jeune Afrique* (13 January 1993).
World Bank. *World Development Report 1992.* New York: Oxford University Press, 1992.
______. *World Development Report 1993.* New York: Oxford University Press, 1993.
Zartman, I. William, ed. *The Political Economy of Morocco.* New York: Praeger, 1987.

5

China: An Emerging Asian Tiger

This chapter examines China's emergence as one of the newest tigers. Anyone traveling to Guangdong Province in southern China in the early 1990s could not miss the dynamic growth in that region. The struggle to provide the infrastructure of roads and telecommunications to keep up with the frantic pace of development reflected the fact that Guangdong, with an annual growth rate of over 20 percent in recent years, is becoming the tiger in the Middle Kingdom—a process that is pulling the rest of China along in a mad scamper of development. Guangdong Province, immediately north of Hong Kong, is booming, being fed a steady diet of foreign investment, government incentives, and skilled entrepreneurs. The push to make money is also stirring other parts of the People's Republic of China, making this nation of over a billion people one of the largest economies in the world. In the 1980-1992 period its real GDP rose by a startling 9.1 percent, the third highest growth rate in the world.[1] In fact, China's rapid expansion could make it the world's biggest economy soon after the year 2010.[2]

The emphasis on private sector-led growth, in both rural and urban areas, is reflected in booming farmer's markets, frenzied speculation in the country's fledgling stock exchanges, and the launching of Chinese stocks on the New York Stock Exchange. Ironically, all of this is in one of the world's last remaining "communist" states. There is more irony when considering that the economy's nonstate sector (including cooperatives, family and individual enterprises, and joint ventures with foreigners) produces about half of the country's industrial product and is continuing to expand.[3]

The clarion calls that China is "arriving" to a higher level of economic development are loud and many. For example, China's credit rating, which affects its ability to access international capital markets, was upgraded in September, 1993 to A3 by Moody's Investor Service, which places it on the same level as Hong Kong and close behind Malaysia and Thailand (rated A1). The country's 12.8 percent GDP growth in 1992, 13.4 percent in 1993, and 11.8 percent in 1994, the highest growth rates anywhere in the world, appeared to confirm that a new tiger is being born. Indeed, the U.S. publication *Businessweek* in 1993 bullishly proclaimed: "From the upscale cafes in the southern city of Guangzhou to the gritty steel mills of the industrial north, China is in a dash for prosperity. Even after Deng Xiaoping passes from the scene, there will be no stopping the momentum."[4]

The rest of the world appears to be rushing in to China and buying all things that the country can offer investors. In 1993 alone, it was expected that contracted foreign investment was to exceed $100 billion, impressive by any standard.[5] The experience of Brilliance China Automotive Holdings, a major bus maker, reflects China's new leap forward. After working with a team of Hong Kong-based Arthur Andersen accountants for several months in 1992, the company was reorganized around its assembly operation and reincorporated in Bermuda. On 9 October 1992, Brilliance China Automotive Holdings made its debut on the New York Stock Exchange, its offer forty times oversubscribed.[6]

Yet another reflection of a changing China is that the rash of investment in the Middle Kingdom is not all in one direction. With Hong Kong's return to Chinese control looming on the horizon in 1997, Chinese companies are combing the British colony for purchases. For example, in 1992–93, the state-owned Shougang Corporation, one of China's largest conglomerates, bought stakes in four Hong Kong companies at the cost of $121 million.[7] The $1.8 billion iron and steel company is also becoming a multinational corporation actively looking overseas to expand its operations. In 1992 and 1993 it spent $120 million for an iron mine in Peru and for an undisclosed amount purchased a Fontana, California steel plant it dismantled and shipped home.[8] Another strategy for the company has been to enter the high-tech business through strategic alliances. Along these lines, Shougang made a $200 million deal with Japan's NEC Corporation to make semiconductors.[9]

The excitement about China is not without its detractors. As *Time Magazine*'s Richard Hornik commented: "China's economic 'boom' is more mirage than miracle, and rosy predictions are based upon its neighbors' successes, not Beijing's ability to sustain growth. The regime is more akin to Latin America's hyperinflationary Peronistas than East Asia's ascetic militants. Beijing has flunked the fundamentals of sound fiscal and monetary policy and proven incapable of accommodating the impulses of a free market. Inflation, speculation and lax regulation are fueling a bubble economy."[10]

Although enormous problems exist—a low per capita income, inadequate infrastructure in many parts of the country, corruption, a legal system in transition, transportation bottlenecks, and political discontent with the authoritarian system—China's economy is undergoing a sweeping overhaul that will take time and clearly put the political system under pressure. As the same time, China's momentum on the economic front is substantial. This newfound economic muscle also has a broader-ranged political component. Many of China's neighbors are increasingly aware of its military and economic power and the role that it plays and will play in Asian affairs.

The People's Republic, Mao, and the Fight Under the Red Flag

The People's Republic of China was founded in 1949 by the Chinese Communist Party (CCP) under the leadership of Mao Zedong. The CCP came to power in the aftermath of the Japanese defeat in the Second World War and a hard-fought civil war (1947-1949) against the Nationalists or Kuoumintang, led by Chiang Kai-shek, who retreated to Taiwan. The CCP rapidly embarked upon recreating China, which meant sweeping away old social mores, eliminating all political forces other than the CCP, and large-scale industrialization. The early model for the new China emerging from the old was Stalinism. As Professor Ranbir Vohra noted, "From 1947 to 1956, when the CCP was dismantling the old order and replacing it with their new system, the Maoist model and the Soviet model were used concurrently."[11] This initially did not provide any major problems as Maoism with its emphasis on being a better "Red" than expert was largely adhered to in the countryside and Stalinism in the urban areas. Mao would eventually diverge from Stalinism. By the late 1950s Mao felt that the country's pace of development could move more quickly if the "mass movement" of the people was tapped. Arguing that point via the Great Leap Forward campaign, launched in 1957, China would make rapid advances—such as matching the United Kingdom's steel production within fifteen years.

The core of Mao's mass movement was the shift of the country's agricultural population—the vast majority of Chinese—to communes. Like the military machine that propelled him and the CCP to power, the communes, with their strictly egalitarian, disciplined work units, would propel China into the ranks of industrial giants by turning the peasants' efforts to steel production in backyard mills. The Chinese economy did not take well to this prescription in the 1957-1959 period; the Great Leap Forward was in reality a Great Leap Backwards.[12] The steel produced failed to meet industrial standards and many of the makeshift furnaces dissolved in rainstorms. More damaging was that the turn to heavy industrialization undermined agricultural production, leading to widespread food shortages and famine by 1960; estimates of the dead from starvation range from 20 to 46 million.[13]

Although the Great Leap Forward was abandoned in 1961 and Mao maintained considerable power, the stewardship of the economy was turned over to Liu Shaoqi and Deng Xiaoping. Despite many challenges, the Liu-Deng team was able to steer the nation into a more productive mode by the mid-1960s. Mao and other CCP radicals, however, regarded Liu and Deng's use of limited market mechanisms in repairing the economy as "capitalist pollution." Mao reemerged from the sidelines in 1965 to launch the Great Proletarian Cultural Revolution, which placed an emphasis on ideology over expertise (i.e., the old "red was better than expert" line).[14] The youthful standard bearers of the movement, the Red Guards, disrupted public life throughout China, brought the economy almost to a standstill, and were responsible for the death of millions. Much of the CCP old guard was swept away and the Gang of Four, led by Mao's wife, Jiang Qing, spearheaded the radical movement with little regard to the consequences for the country.

Political instability continued into the 1970s. A growing rivalry between Mao and Marshal Lin Biao, head of the armed forces and once designated to succeed the Chairman, led Lin Biao to plot a coup. An assassination attempt was made on the Chairman's life, but the coup attempt failed and Lin Biao was killed in 1971 in a plane crash in Mongolia, when he was fleeing the country. Out of this chaos, Zhou Enlai emerged as the most powerful figure in China. Zhou restored Deng Xiaoping to power in 1973 to help put the economy on track and in 1975 launched China's new economic program, the Four Modernizations. The new program placed an emphasis on the modernization of the Chinese economy through an upgrading of the country's agriculture, industry, defense, and science and technology. The political atmosphere, however, remained unsettled and on the death of Zhou in early 1976 the radicals, led by Jiang Qing and the Gang of Four, once again purged Deng. When Mao died in September 1976, China appeared on the brink of political instability.

Within China, the population was weary of decades of political turmoil and the power behind the throne, Jiang Qing and the "Gang of Four," promised more of the same. Although relations had improved with the United States, there had been troop clashes with the Soviets in the late 1960s and the atmosphere remained tense along the long Sino-Soviet border. As Anne Thurston notes: "At the time of Mao Zedong's death in September 1976, the People's Republic of China stood precariously at the brink—the populace demoralized by decades of class struggle and political persecutions, the economy crippled by socialist constraints, the more competent of the Communist Party's leaders either purged or dead, the government intimidated into inaction by its aging dictator-emperor and the radical clique that surrounded him. The communist regime, it appeared, was losing all legitimacy—or what was once called the Mandate of Heaven."[15]

Chinese dynastic history is defined by a cyclical pattern in which new dynasties rose when old dynasties become corrupt, military enfeebled, and administratively ineffective. Whenever the imperial order declined, internal rebellions arose, and challengers arrived to ascend the dragon throne. In some cases, the aging dynasty rose to the challenge, rooted out corruption, revived the moral codes upon which it was founded, and embarked upon a renewal. In most cases, the dynasty failed to defeat new challengers and keep a new dynasty from emerging.

By Mao's death in 1976, China hung in the balance between the death of the Communist dynasty and regeneration. Not only did Mao's successor as Chairman, Hua Guofeng, lack a strong base of political support, Jiang Qing and the Gang of Four appeared ready to plunge China into another round of ideological struggle. The Communist regime, however, survived because the CCP's more moderate elements rallied around Deng Xiaoping, and the radicals were soon under arrest and their power broken. Hua was quietly shunted aside. In this environment of political uncertainty, Deng gradually assumed greater power until 1978, when he was the paramount leader. Having gained power, Deng began to reshape

China in his mold—which was very different from Mao's. Regardless of the difference, Deng gave a new lease of life to the CCP dynasty.[16]

Putting the Tiger Factors to Work—Round One

What was Deng's vision for China? He was, after all, a survivor of the Long March and a lifelong communist. Unlike Mao, however, Deng was a pragmatist and not bound by ideology. It was he, after all, who had stated at a party congress in the 1960s when propounding the use of market mechanism to stimulate economic activity: "It does not matter whether the cat is black or white. So long as it catches the mouse it is a good cat." Having witnessed firsthand the failures of such developmental efforts as collectivization of farming and the Great Leap Forward, his desire was to use capitalist tools to convert China into a modern, industrialized nation-state. This entailed both strong industrial expansion as well as an improvement in the standard of living for the population in general, but on a carefully managed, sequenced program in which the party did not lose control. Equally important, Deng's reforms helped create constituencies in the population (such as farmers, party members, and the military) who benefitted from and came to identify with the economic reforms, an important development in maintaining a political base. Hence, under Deng's leadership a number of tiger factors became manifest.

Maximizing Agriculture

One of the first areas liberalized was farming, which allowed China to maximize its agricultural productivity. By 1978, China, despite its large agricultural sector, was no longer self-sufficient in grain due to political turmoil and was forced to import grain to supply about 40 percent of its urban population.[17] The key elements of Deng's agricultural reforms launched in 1979 were a marked reduction in the interventionist role of state planning and the demobilization of the commune system. In a gradualistic fashion, the peasants were given commune-owned land and their obligations to the collective were lessened considerably. Although theoretically, allocated land could not be sold, the peasants in many respects came to acquire real control over production. By 1984, a new phase of rural reform began that, under State Council Document No. 4, formally abolished the system of the "people's communes," replacing them with township governments. The result was that agricultural production rose, at the rate of 3 percent a year.[18] Although the point should not be overemphasized, this change meant that in many rural areas the standard of living dramatically increased and a new group of better-off peasant farmers emerged.

Another significant rural reform was the creation of township and village enterprises, or TVEs. Owned by local governments, the TVEs offered incentives to Chinese industry to relocate or start-up operation in rural areas where regulations were relaxed. The TVEs were highly successful. When Deng came to power in 1978 such industry accounted for no more than 20 percent of rural output; in the early 1990s it accounted for some 45 percent, while 40 percent of the country's industrial laborers lived and worked in rural areas.[19]

Foreign Investment Welcome

One of China's strongest tiger factors is its openness to foreign investment. This was not always the case as the Mao period was often highly xenophobic and demonstrated a preference for economic self-sufficiency. However, Deng recognized the importance of foreign investment and in 1980 established the Statute of Joint Venture and the new Commission for Foreign Investment. The impact of these measures was significant in shaping China's future. Paul Theroux notes, "They meant that foreign companies could start factories in China with Chinese partners, that capital and technology could be easily transferred, and perhaps most important, that foreign loans (both government and private) were permitted."[20] Moreover, China's Special Economic Zones were created with the intention of attracting foreign investors to China's cheap and willing labor pool.

China's effort to welcome foreign investment certainly contrasts with the disappointing record of the Philippines. Unlike China, the Philippines pursued policies in the late 1980s and 1990s that made it less attractive to foreign investors. China offered cheap labor and relative political stability as it actively sought foreign investment. In contrast, the Philippine government was content to vote down a continued U.S. presence in Subic Bay, depriving itself of several billions of dollars in rent, not to mention sending out the message (as it was perceived by a number of U.S. firms) that U.S. investment was not welcome. Considering the highly competitive international environment for attracting international business, China's measures have proven to be highly successful.

Trade Liberalization

Deng also opened China to foreign trade and foreign investment through joint ventures and joint management. In November, 1987 CCP General Secretary Zhao Ziyang called on China's coastal areas to follow an export-led growth strategy. In 1988, the government, especially the faction of the CCP supported by Deng under the leadership of Zhao, promoted rural enterprises as opposed to the stagnant state-run economic units, to get involved in foreign trade and urged state-owned firms to take the flexible management styles of rural firms as a model.[21]

The new economic environment was quickly taken advantage of in southern China. Guangdong Province, in particular, was able to benefit from its proximity to Hong Kong and make major strides in its economic development. In the 1990-91 period, the province's growth in exports was 18.1 percent and growth in fixed income asset investment was 25.8 percent. The key tiger factors at work included maximization of food production, foreign investment (much of it from Hong Kong and Taiwan), and foreign trade. Guangdong's population also enjoyed a reputation for being hardworking and its labor costs were relatively cheap compared to South Korea, Singapore, and Malaysia. Moreover, they proved willing to learn new skills that pushed added value in production up the technology scale. The result of all this was since 1980, Guangdong's real GDP has grown at an average annual rate of 13.5 percent and it boasts of an urban population with a per capita income among the highest in China. Guangdong's success soon spilled over into a number of coastal provinces that found export-led growth a marked improvement over closed frontiers.

Advances made in trade liberalization, supported by price liberalization, were cost efficient for the Chinese government because China's rapid pace of development in the 1980s was largely fueled by foreign investment. That investment was direct, going into factories that created jobs for the local population. Related to this was the Chinese move to production of the same cheap goods that had earlier been made by South Korea and Taiwan. While much of what was produced went for the export market, a significant portion went to local markets as these goods also proved were affordable to the Chinese population.

China's trade liberalization accomplished many goals: it opened the country up to foreign investment and trade, helped create jobs and develop a local class of people who identified with the changes, and reduced the need to turn to foreign banks for loans. China in the 1980s had a small external debt relative to the size of its GDP and compared to other developing countries, such as Argentina, Brazil, and the Philippines, all of which experienced considerable problems in meeting their repayment obligations.

Balancing Economic Reform and Politics— The Road To Tiananmen

China's economic boom was rapid in the 1980s, but it also had an unsettling side effect: the country was confronted with double-digit inflation, which sparked panic buying in major urban areas in the fall of 1988. Corruption, especially in the south, emerged as a major problem. The growing unease among the CCP's hardliners that the economy was lurching out of control resulted in a conservative backlash. The central planners took over control of the economy from Zhao and his reformist clique, and carved out a wide-ranging program of economic retrenchment in late 1988. However, the ability of the conservative politicians to re-establish central control over the country economy was weak (see table 5.1) and besides, new forces had emerged in the political arena to challenge the CCP's authoritarian rule.

China's economic development since 1978 has been marked by flexibility and liberalization, whereas its political development toward a more open society had lagged far behind. Foreign investment, technology, and management expertise had been allowed to pour in and was welcomed; Western liberal ideas, such as freedom of speech and democracy, on the other hand, were resisted and considered, particularly by the authorities, as unwelcome.

The challenge of maintaining political control while granting economic freedoms placed Deng, and other centrists, in a difficult position: the liberals within the CCP and in society in general were increasingly critical of the regime's authoritarian governing style, and the conservative CCP hardliners were wary of economic liberalization. The liberals

TABLE 5.1
China's Inflation

1980	1981	1982	1983	1984	1985
6.0%	2.4%	1.9%	1.5%	2.8%	8.8%
1986	**1987**	**1988**	**1989**	**1990**	
6.0%	7.3%	18.5%	17.8%	2.1%	

Sources: State Statistical Bureau 1992 Yearbook, 207.

favored the ongoing marketization of the economy and would have liked to see a parallel liberalization of politics, including greater powers for the National People's Congress.

Premier Li Peng and the retired, but still-influential economist Chen Yu, representing the hardline faction, favored a more centrally planned economy and limited economic freedom. The hardliners understood the difficulty of opening the door economically while keeping it closed politically. While they grudgingly acknowledged that Western ideas were responsible for the stunning upturn in China's economic life, they were highly apprehensive, and correctly so, that Western political values posed a direct challenge to their authority.[22] The point of difficulty was in determining where economic freedoms should end and political freedoms begin.

The liberal-hardliner collision came in 1989. China had concluded a decade of substantial economic growth—a dizzying pace of 9.7 percent from 1980–1989.[23] More Chinese had been exposed to the outside world than at any time under previous Communist rule; the combination of foreign investment, tourism, new communications, travel, and dynamic economic growth began to raise questions in certain circles about the nature of the political system. Moreover, in Europe the Soviet bloc was rapidly crumbling before a wave of democratic revolution. To many young Chinese students, the awkward issue of democracy was something that was attractive and should be pursued. In the spring of 1989, staggered by the death of the liberal leader Hu Yaobang, hundreds of thousands of students and others poured into Beijing's Tiananmen Square and started to demand greater political freedoms. To the hardliners, the worst had occurred—democratic rebellion was in the air, threatening their positions.

Demonstrations in Tiananmen Square caused a crisis in the CCP's leadership and several weeks (when the government appeared to be totally paralyzed) passed before the hardliners finally outmaneuvered the centrists, purged the liberals (such as party chief Zhao Ziyang), and sent the military into Beijing to crush the student movement. Deng sided with the hardliners. In the days that followed and with the international media watching, the hardliners emerged as a Stalinlike clique ready and able to ruthlessly crush the dissidents. The leadership claimed that Tiananmen was the work of a small group of counterrevolutionaries who advocated bourgeois democracy and the overthrow of the CCP regime.

Not surprisingly, these developments stalled the economic reform process. While inflation was brought down to 2.1 by 1990, economic growth slowed from 11 percent in 1988 to 4 percent in 1990. Despite the international outrage about the Chinese government's handling of events in Tiananmen Square and a sharp reduction in access to international credit (mainly from the West), the hardliners gained considerable control over their country by 1990. While Deng was not ousted from power, his position was temporarily weakened and his ideas of marketization of the economy took a back seat to the party's main objective of staying in power. The struggle over China's future direction, however, was not over.

Tiger Factors—Round Two

In January, 1992, after a long spell of absence from the political scene, Deng returned by visiting the special economic zone of Shenzhen in Goungdong province, due north of Hong Kong. Shenzhen's economic development in the 1980s had been nothing but phenomenal. Together with China's president, Yang Shangkun (one of the key figures behind the military crackdown), Deng made his visit a major affair. He praised Shenzhen's progress and predicted that southern China would soon become Asia's "fifth dragon," joining the ranks of Hong Kong, Singapore, Taiwan, and South Korea. Deng then visited joint venture factories and high-tech enterprises in Zhuhai, another booming special economic zone in Goungdong province. In a calculated action, he lauded southern

China's economic miracle and criticized leftist hardliners opposed to liberalization, calling on the rest of the country to copy the south.[24]

Although the hardliners sought to keep Deng's comments out of the press, they failed to do so in Hong Kong and Guangdong and ultimately in March and April, 1992 the official press on the mainland began to beat to the new drum. This was due in large part to the resurgence of Deng's power, especially at the 20 March 1992 National People's Congress, in which the CCP agreed to "speed up reform and opening up" of the economy. For many Chinese, who had become cautious after the military crackdown in Beijing, the change in the official press (in particular the *People's Daily*) signalled that the economic reform process was on again. As Thurston comments: "The pall that had hung over much of the country since the Beijing massacre began to lift. The mood in China changed. Since then, the race to get rich has been on."[25]
Deng's guiding slogan for this round of implementing tiger factors has been "Develop the Socialist Market Economy," a vague halfway house rhetoric that seeks to bridge the socialist past of Marxism-Leninism-Maoism with international capitalism, and to which the Chinese have actively responded in turn by making more money than ever before. The message in post-1989 China has become clear: the government will leave you alone in making money, but will not tolerate political dissent. The overwhelming trend today is to make as much money as one can and leave politics alone.

The key elements of the second round of economic reform are a deepening of trade reforms, financial reform (in particular the banking sector), and development of the infrastructure. Other areas in the second round of reform are an overhaul of the national tax structure, the revitalization of Shanghai as a major regional financial center, and measures to maintain China's attraction for foreign investment. The objectives of transforming China into a major world economic power are ambitious, but the Chinese leadership continues to move ahead in a gradual manner. Above all else, the new round of economic reforms will be carefully sequenced.

Foreign Investment

Despite a slowdown following Tiananmen Square, foreign investment in China gradually returned. In 1992 alone, foreign investors put $11.2

billion into the Asian nation and signed agreements for $57.5 billion of future investments.[26] China's policy of welcoming foreign investment entered a new phase when it was announced in 1993 that additional fields were opening on a "trial basis," including finance, insurance, retail businesses, and construction.[27]

One of the important dimensions of the foreign investment regime in China is its pragmatic nature. Although the People's Republic and the government in Taiwan officially have no diplomatic relations, Hong Kong has provided a welcome conduit for Taiwanese money into mainland investment projects, especially in southern China. Moreover, Hong Kong is geographically closer to Taiwan than anywhere else, with the exception of the PRC. This has meant that a complex interdependence has evolved.

At the heart this interdependence is trade and investment. In the 1980s the cost of Taiwanese labor gradually became more expensive, and it was useful to subcontract work through Hong Kong to the lower-waged mainland China labor pools. For its part, the government in Beijing was interested in the trade and investment benefits accrued by such foreign operations. Moreover, in Beijing's perception, the growing interdependence with Taiwan would eventually create a convergence in development that would ease the process of unity. As Taiwanese investment flowed through Hong Kong into south China, important links were established that bound the three regions together.

The actual amount of capital involved in the Taiwan-Hong Kong-south China triangle is substantial and growing. In 1990, Taiwan's indirect exports to China via Hong Kong totalled $3.3 billion; by 1992 that amount was estimated to have leaped to $7 billion.[28] This meant that exports to China accounted for 6 percent of Taiwan's total exports and 2 percent of its imports—not bad for two states that claim to have no official relations. Additionally, by 1992 Taiwanese investment in China was estimated at $5 billion, most of which was in Fujian and Guangdong.[29] Taiwanese investment continued to be a factor in 1993 and 1994.

Although the Taiwan-China relationship is different from that between Beijing and Hong Kong, it is probable that Taiwan, with one of the world's largest international reserves (over $80 billion), will continue to be a factor in China's economic development. Short of a return to outright hostilities between Beijing and Taipei, both sides will gain economically maintaining this strategic and increasingly blatant link. In fact, discussions between the two governments in 1994 on a number of issues revolving around trade and fishing rights indicates that both sides quietly acknowledge each other's existence.

Infrastructure and the Tech Factor

To match the lofty goals of being one of the world's largest economies, China has embarked upon a number of major infrastructure projects. At various stages of planning or construction these include: the Beijing-Kowloon Railway, a $3.3 billion project that entails the construction of 549 bridges and has 200,000 workers; the Guangzhou-Shenzhen Expressway, a $1.2 billion highway under construction that will speed travel within southern China; the Nanning Kunming Railway, a $1.1 billion line that will help develop the coal-mining industry; the Three Gorges Dam, a $10 billion project on the Yangtze River; and the development of the Pudong industrial and commercial zone, which includes an airport and subway and has already attracted $10 billion in investments.[30]

Another component of infrastructure for China is an upgrading of technology, particularly telecommunications. China lacks adequate telephone lines and public switching equipment. Without an ability to "reach out and touch someone," Chinese exporters run the risk of producing goods that may not come to market or the right market at the right time. Additionally, the economic transformation occurring in China may miss a vast domestic market.

The Chinese government is well aware of the need to bridge the gap in communications. Demand, however, is relentless—China already has 17 million phone lines and its has almost reached its target for the country's five-year plan, which ends in 1995.[31] The goal is to have 100 million lines operating by the year 2000. The pressing need to install, upgrade, and expand China's telecommunications has meant the import of foreign technology. One such area that is booming is cellular phones. Although subscribers to cellular phone services were under 300,000 in 1993, the market is growing dramatically. Simply stated, the cellular phone is becoming an important tool in China's development. It is no coincidence that of the 300,000 cellular phone subscribers in China, the booming Guangdong Province has more than 200,000.[32]

China is also actively involved in the development of its satellite communications. In early 1994, it agreed to launch eleven telecommunications satellites for the U.S.-based Hugher Communications over the next thirteen years. The China Great Wall Industry Corporation will be responsible from the Chinese side, helping in the launches from Sichuan Province.[33]

Financial Reform

Reform of China's financial sector has lagged, but in 1993 and 1994 it became the focus of official attention. Because of the transitional nature of the economy, China's financial system entered the 1990s in a state of upheaval. This has meant that China's banks and stock markets have operated in periods of old, outmoded rules, no rules, new rules, and poorly understood and weakly enforced rules. Financial authorities and interested outside parties have been confused by all of the above. This situation made financial reform a necessity.

The country's two stock markets, in Shanghai and Shenzhen, enjoyed a remarkable boom in the early 1990s. As many Chinese and foreigners discovered these stock markets, a substantial spurt of investment occurred. However, there was a downside. In most respects, China's two official stock markets were, according to *The Financial Times* (25 March 1993), "unruly and virtually unregulated." In 1992, trading volume in A shares (for local Chinese) and B shares denominated in U.S. dollars (for foreigners on the two exchanges) exceeded $17.2 billion.[34]

Shares of local Chinese companies were also available outside of the official stock markets, where the buying and selling of shares occurs with no legal safeguards or guarantees. Most sellers of shares are cash-starved companies who are denied bank loans or listings on stock exchanges. Unregulated stock trading poses a strong challenge to Beijing's ability to coordinate the country's financial system, especially if thousands of companies coax household savings into unauthorized shares.[35] Additionally, illegal trading could retard the development of securities regulations, brokerage firms, and institutional investors. These factors are critical links in the development chain for China to pass from an underdeveloped nation into a new tiger. For the unwary investor—foreign as well as Chinese—freewheeling market mechanisms without any controls are likely to spell disaster.

China's banking sector was described by *The Economist* in early 1993 as a "mess." State controls in China's banking sector are far more dominant than in many of its neighbors, such as South Korea; all banks are state owned and politically subservient. Until the mid-1980s, the People's Bank of China, the central bank, was responsible for all commercial bank activity in the country. Eventually four "specialized banks" were created to handle different aspects of commercial banking, while the remaining banks that exist are expected to make a profit.

While numerous sectors of the economy have experienced strong growth, the banking sector has lagged. As the experience in Central and Eastern Europe exemplified, the accumulation of bad debts on state banks' books is done to prop up loss-making and uncreditworthy state-owned enterprises, which eventually becomes a liability to the forward momentum of the economy. In China, some of the negatives of these practices were observable in the early 1990s. As *The Economist* (27 March 1993) observed: "[I]t is almost impossible for the central bank to control the money supply. Since the branches of both the central and specialized banks are treated as administrative units of government, local officials have considerable power to ignore orders from the central bank to tighten money supply."

The flaws in the Chinese banking sector became evident in 1993 when farmers rioted in Sichuan and Guangdong provinces. The root of farmer grievances was the banks' failure to provide the expected state payments. The farmers are required to sell their produce to the state, but instead of receiving cash, the farmers received IOUs from the banks, which are not honored for long periods of time, hence the rioting.

By 1993 a number of measures were taken by the government to restore discipline. Under the new laws, banks were directed to cut links with property development units. This came after a scandal when officials at the Agricultural Bank of China were dismissed for diverting large sums to property speculation. In mid-1993, the new economic czar, Zhu Rongji (also Deputy Prime Minister and Governor of the Central Bank) was installed to impose discipline on the banks and the credit system in general. New measures were introduced that included the creation of a more autonomous central bank (with monetary controls), plans to convert the existing state banks into commercial banks, creation of a new set of state development banks, and the guarantee of a reliable flow of tax revenues for investment in infrastructure of national significance.

The banking sector received further attention at the Third Plenum of the Fourteenth National People's Republic Congress in 1994 that passed a series of laws to adapt the financial, tax, accounting, and legal systems to a market economy. In particular, legislation was passed to commercialize state banks, create specific "policy" banks, and strengthen the central bank. Under the new reforms the four state banks—the Bank of China, the Agricultural Bank of China, the People's Construction Bank of China, and the Industrial and Commercial Bank of China—will no longer disperse policy loans. Additionally, as accounting and management reforms are implemented, the four state banks will become responsible for their own profit and loss. The policy banks that were created were the State Development Bank, the Agricultural Development Bank, and the China Export-Import Bank.

Financial reform is slowly emerging as a tiger factor for China. Although progress has been slow, measures being taken are carefully measured and China's ruling elite has no desire to be whipsawed by shock programs that have characterized much of Central and Eastern Europe's programs. Reforms in 1993–1995 will help establish a more unified financial system by expanding the capacity of the money and capital markets and enhancing supervision of banks. Additionally, reform of the central bank will help consolidate central authority in banking affairs and break down regional separation. As one investment bank study noted of the prospects for financial reform: "The learning curve for the new generation of Chinese bankers and bank supervisors is steep, but not impossible. China's authorities realize that banking reform is critical to guaranteeing the ongoing pace of marketization."[36]

Revitalizing Shanghai as a Financial Center

Another dimension of China's new leap forward is the revival of Shanghai as a financial center. Although the Shanghai Stock Exchange (SSE) is currently light years behind Tokyo, Hong Kong, and Singapore, it was once a paramount center in East Asia and plans have been activated to regain some of the past glory. According to *Euromoney,* the SSE will have outgrown its nursery in the gallery of a former hotel ballroom by 1995 and be moved to a new building in a development zone known as Pudong, where it is expected to have 3,000 seats.[37] Foreign brokerages will also be included. The bullishness about Shanghai's prospects was articulated by the Industrial Bank of Japan's Eijo Daimon: "It is very realistic that it will become more important than Hongkong. In five to 10 years, there is no doubt Shanghai will be an international financial centre."[38]

Shanghai's market capitalization at the end of February, 1993 was $21 billion compared to Hong Kong's $197 billion, indicating the vast distance that Shanghai needs to travel to regain its major financial center status.[39] The Chinese have high hopes for the city's return to glory and regard Shanghai's revival as related to Hong Kong's reversion to China in 1997: after that date Hong Kong will dominate Guangdong and South China, while Shanghai will attend to the financial needs to the rest of the country. Zheng Bailin, general manager of the Bank of China in Shanghai, stated of the relationship between the two cities: "China is a vast country. They will help each other. After 15 to 20 years Shanghai will be a very important economic and financial centre, not just for China, but for Asia and the world."[40] A number of Japanese, U.S., and European investment banks have already established representation in Shanghai and more brokerages and investment houses are expected.

Shanghai's revival is not limited to finance, though other economic activities reinforce the drive to recapture its status as a financial center. Heavy engineering, textiles, food processing, and machinery are all found in the city or its suburbs, which have a population of 7.86 million in the city proper and 5 million in the surrounding rural areas under its jurisdiction. Statistically, Shanghai's boom is reflected by its accounting for 7 percent of China's industrial output and 15 percent of its trade in 1991, while its economy expanded by 14.8 percent in 1993 followed by first half growth in 1994 of 13.6 percent.[41] As one Western diplomat noted of Shanghai's revival: "They are born capitalists. If the open door continues, their innovative ability will come to the fore. They are the same people who built Singapore, Hongkong and Taiwan."[42]

Challenges

China's economic development is impressive, yet there are a number of challenges. Growing problems in China include high inflation, widening class differences, crime and corruption, the need to overhaul the tax structure, and transportation bottlenecks. The weight of these problems should not be overstated, nor should they be understated. While it is dubious that the reform process will be halted, each of these problems challenges the pace of change.

Inflation

One of the major areas of concern for China's economic policymakers is inflation. In the late 1980s, economic reforms resulted in strong growth as well as high inflation, partially due to price decontrols. Inflation in turn fueled public concern about rising prices, especially for foodstuffs, and resulted in a brief return of hardliner control of the economy. Although inflation was brought down by tight money policies, it returned when the pace of economic reforms picked up again in 1991 and 1992 (to 6.9 percent). China's high inflation is largely caused by the country's rapid growth rate. The faster the real GDP growth rate, the greater the inflationary pressures.

China's inflationary problem is compounded by the central government's difficulty in regaining control of monetary policy due to a decentralization of powers, including regional banking powers. Although China's inflation rate of 24 percent in 1994 (year end) should not be overstated as a major problem, inflation can become a problem if corrective measures are not taken at some juncture to reign in prices. Much of this will depend on the power relationship between Beijing and the provinces. In the third quarter of 1993, some progress was made in slowing the pace of economic growth and reducing the pressure on inflation, but by the fourth quarter inflation was back. It surged again in 1994 and appeared to decline marginally in early 1995, but remained high.

Widening Class Differences

Rapid economic development represents a challenge of balancing economic freedoms with the maintenance of wide-ranging benefits of the capitalist experiment for the vast majority of the population. Without a certain sense of balance the experiment runs the risk of creating a very wealthy, yet small class of entrepreneurs, and a large impoverished class of urban and rural workers—much as it was before the revolution.

The 1993 farmers' riots mirrored growing differences between those plugged into the new economy and those who are not. The focus on industry over the last decade at the expense of agriculture has changed the earlier path of development that favored the farmers. While industrial output expanded by 19 percent in 1992, agricultural production growth was only 3 percent.[43] The imbalance is also reflected in the standard of living. While farmers' annual net income has jumped almost sixfold since 1978 to 770 yuan ($135) in 1993, it is only about 40 percent of urban residents' income.[44] This situation has bred resentment and frustration about the government's policies. It has also created growing numbers of rootless people who have drifted from the countryside to the cities in search of work. Although the standard of living has increased markedly since 1949, the growing disparity between rich and poor, oftentimes urban and rural, is a potentially dangerous development that could undermine China's attainment of tiger status.

Crime and Corruption

Crime and corruption are two more potential obstacles that could undermine China's rush to become a modern industrial society.[45] Under conditions that grant permission to government officials to go into business, there have been cases where some have taken advantage of China's transitional economic situation to abuse their power and position. By the summer of 1993 President Jiang Zemin was forced to acknowledge that "corruption is a virus that has infected the party's healthy body. If we just ignore this phenomenon, it will bring down our party and our system."[46]

One of the reasons given for the weakness of the Chinese currency in 1993 was related to corruption. Despite China's trade surplus and record foreign investment, the local currency, the *remembi,* remained weak vis-à-vis the dollar. According to one source, this was because the government officials running large state enterprises as well as many in the private business enterprises are moving large funds overseas as they prefer to keep their money outside of China.[47] One estimate in 1993 put the amount of money flowing abroad at $28 billion, some of that probably going into illegal activities. Chinese government officials noted that more 21,000 cases of smuggling, counterfeiting, and fraud were solved in the last couple of years.[48] The fact that Chinese money is flowing out of the country raises questions about the nature of the investment environment, especially considering that foreign funds are pouring in and economic growth continues to be dynamic.

China is also witnessing a resurgence of kidnapping, prostitution, and drug trafficking. Reports from Yunnan and Guangdong provinces indicate that drug abuse is on the rise and that trafficking groups are active.[49] This situation does little to reinforce the public's faith in political systems' ability to establish a clear direction. As political scientist Richard Baum commented on the impact of rising crime and corruption in China's social fabric, "Up and down, the bustling, densely populated east coast of China, from Harbin to Hainon, observers noted a deepening mood of alienation and anomie among young people, large numbers of whom were increasingly turning to 'get-rich-quick' schemes, to religion, to the martial arts, or to sex, drugs, and rock-and-roll as quick-fix antidotes to decaying belief systems, declining moral standards, and disintegrating social controls."[50]

Taxes

Another significant challenge for China's government is the pressing need to overhaul the tax system. China's economic reforms have stressed decentralization from the political center to the provinces in order to facilitate decision making. Other measures included a reduction in taxes. While this has been highly successful in stimulating regional growth, it has increasingly left the central government with an ever smaller share of tax revenues. In the past sixteen years, the central government has watched its tax take as a share of GDP halved.[51] As William Overholt notes, "As it is, the central government delegated so much tax authority to the provinces that the central government is inadequately funded."[52]

By the time of the central committee meetings in 1993, it was evident that the center required more revenues—something that would not gain easy support from the provinces. In fact, earlier efforts to redistribute the tax burden were defeated by the provinces. However, the tax question is critical to the central government, which is often hard pressed to appropriate capital necessary for major power and transportation projects.

Prior to 1994 the tax structure in place allowed provincial governments to keep any profits exceeding the amounts they have agreed to remit to the central government. What evolved was a system in which the most successful provinces continued to remit an increasingly smaller part of the pie, while less prosperous provinces have struggled to meet their quota. In 1994, Chinese authorities at the third plenum of the Fourteenth National People's Congress moved to implement a new system in which enterprises in the provinces would pay taxes to both central and local governments based on a scale that is uniform across the country.[53] There were four characteristics to this new tax system: a National Tax Service was created to collect most taxes; enterprise taxes were reduced from 32 to 18; a restructured and uniform Value Added Tax was adopted; and personal income taxes were restructured. This system is expected to augment the central government's revenues and bring them in line with obligations.

There is always a difference between theory and practice, which reflects the tax issue in China. For China's economic success to continue and for it to avoid centrifugal local forces, the ability of the central government to finance itself without running up a large domestic or external debt is important. This remains a challenge.[54]

Breaking the Iron Rice Bowl

China's movement to privatize its state-owned enterprises (SOEs) has been gradual and for good reasons. Many of China's large public corporations have provided not only employment, but a plethora of social services, including schools, infant care, and hospitals. In 1993, the 108,000 SOEs employed a little over 110 million workers including dependents; however, the state enterprise system has under its umbrella about 340 million people.[55] Consequently, China's major state-owned companies have provided what has been referred to as the "iron rice bowl."

As China moves to restructure its major corporations to become more attractive to foreign equity, elements of the iron rice bowl are being stripped away. The corporatization of China's major companies—the shift to more market-oriented practices—raises many social questions. Most significantly, who will provide those services formerly supplied by the steel works or the local brewery? This obviously carries political as well as financial implications. The challenge of reforming the SOEs will not be easy considering social and political factors. As Paul Schulte noted: "Given the absolute size and enormous contribution of these enterprises to economic output, the reform of SOEs will be a mighty undertaking. It will take a very long time. It will not happen without some pain."[56] This is one area where the clock is ticking for China in terms of the need to reform.

Political Questions

As in the case of all authoritarian regimes, hard or soft, the political question of sequencing political liberalization remains a factor that cannot be ignored. To this are attached the ideas of legitimacy of the country's political institutions and the rule of law. This is especially the case as economic freedoms move apace, but political freedoms lag. Moreover, the nature of the economic freedoms are freewheeling and, in a sense, promote the idea of challenging the old, authoritarian order. Another layer of tensions is that between those parts of China that are achieving breakneck economic success and those that grow at a slower pace. Although the potential for civil war resulting from regional or class differences or between liberals and conservatives is not high in the medium term, the potential for a systemic upheaval in China cannot be ruled out. As David Shambaugh, a professor of Chinese politics at the University of London noted in July 1993: "China is at a key transitional point right now. A watershed is coming. The center is challenged by a variety of social frustrations. Once they are triggered, there will be very quick snowballing of popular protest and the state will be hard-pressed to cope with them."[57] The farmers' riots in mid-1993 were one indication of the potential for political instability. The ultimate risk is that China could break into a number of regions like the former Soviet Union or give rise to a military dictatorship. The problem with China's economic modernization is that it contributed to the decentralization of decision-making, which diluted central authority to control such areas as monetary policy essential for maintaining lower inflation. Consequently efforts had been made in 1993 and 1994 to regain control of macroeconomic policy.

Simon Winchester's futuristic novel, *Pacific Nightmare: A Third World War in the Far East* (1992) centers around the turnover of Hong Kong to the People's Republic in 1997: Having survived the succession of Deng, Beijing places a Maoist hardliner in the governorship of Hong Kong when the British depart. This is done because of growing concerns in the Chinese leadership about the "polluting" impact of Hong Kong on the rest of China; Its freewheeling economic dynamism and limited political liberalism threaten Beijing's control over the entire nation, which has continued to embrace the market. Winchester argues, in his fast-paced thriller, that the economic boom in China has underscored strong differences between the more doctrinaire north and the more capitalist south. He states that " in general the provinces of southern China were fast gaining a solid appreciation of the world outside, and of the possibilities that could be open to their own people. South of the Yangtze River there was indeed a 'new' China developing—but one that was evolving not, as Party cadres once had hoped, into an immense ideological communion, but into an ethnic whole that was united by a common hope for modernism and prosperity and individual freedom."[58]

Winchester is echoed by Yale University's Helen F. Siu, a professor of anthropology, who noted that cultural identification with China's political center is highly diffused with social life and that the south has "in the hierarchical territorial map of power holders in Beijing, has always been one of the most distant from the center."[59] To northerners the south was for centuries a place where disgraced imperial bureaucrats were sent and "civilization" was a thin veneer. Throughout most of the PRC's existence the south has also been closely associated with the dangers of "spiritual pollution" threatened by Hong Kong's strong capitalist bent. As Siu concludes, "One expects that south China has the potential to break away."[60]

For all those that are apprehensive about China's ability to deal with the difficult question of political reform—which would include legal reforms—and perceive this as a critical weakness, there are those who regard Beijing's authoritarian nature a plus factor. Barton Biggs, chairman of Morgan Stanley Asset Management, noted in 1993: "I am especially bullish on China because it recently devalued its currency by 20%, getting rid of an artificial official exchange rate and going to a market exchange rate. Their economy is going to be open and deregulated."[61] To this he adds that when comparing investment prospects for China and India, "In the short run the potential is greater in China because it is a dictatorship, while India is a coalition democracy."

What Lies Ahead?

Whatever may be one's perspective, the political factor remains one of the most unpredictable elements in projecting China's future. However, it is likely that China will follow the path of South Korea and Taiwan—a gradualistic move toward a more open political system, occasionally interrupted by crackdowns on the opposition. Although tensions will rise when Deng passes from the scene, it is probable that someone within the current national elite will emerge to continue the reform process and maintain China's unity. Indeed, it was speculated that in 1994 the country's leadership, including President Jiang Zemin, Premier Li Ping, and Finance Minister Zhu Rongji, had already shifted to a more collective form, considering Deng's advanced age.[62] Hong Kong's move to Chinese control adds another political wild card, but one which can be controlled and most likely channelled in the right direction in cementing national unity. An important underlying factor in maintaining China's forward momentum is the improvement in the population's standard of living and that despite the lack of political freedoms, the population perceives that the leadership will continue to improve on its lifestyle.

Central to these developments is the question of whether a civic society can be formed in China. Clearly in the 1990s, the viability of Marxism-Leninism-Maoism as a ideological core has faded badly and China is drifting without a well-defined sense of direction beyond Deng's somewhat confusing slogan to "Develop the Socialist Market Economy." Perry Link, a professor of East Asian Studies at Princeton University, captured the dilemma: "A fundamental question for China today is: What values and institutions can help to restructure a civil society within the current vacuum?"[63] Confucianism has been discredited, which weakens but does not entirely discount that option. The remaining path is that of making money—a task to which the nation appears to have adjusted while putting the political questions on hold.

What is evolving in China in the 1990s is what *New York Times* former chief of the Beijing bureau, Nicholas D. Kristof, referred to as "Market-Leninism"—a combination of a highly disciplined, one-party rule with centralized decision making, supported by a dynamic market economy. Accordingly, the reformists plan to jettison communism—but not Communist party rule—and move China into the East Asian tradition of free-market soft authoritarianism, a path already trod by South Korea and Taiwan in the 1960s and 1970s. In the cases of South Korea and Taiwan, the authoritarianism has often been attributed as a major factor (along with others; see chapters 1 and 2) in the successful transformation of those nations from troubled Third World countries into tigers. As Kristof commented, "In short, dissidents are zapped with cattle prods and the economy is zapped with market incentives."[64]

While it is questionable that Market-Leninism will entirely fill the ideological core, it likely that the economic forces shaping the country will create enough of a constituency that regards the need for stability as critical. This will guarantee that the progress made in converting China from a totalitarian state to an authoritarian state will last and that the market remains central to China's growing global stature. A return to ultraleftist Maoism is not a probability, considering the memories of the disruption caused by the Cultural Revolution. As Overholt aptly notes, "All the most powerful political groups in China see a stark contrast between the successes of gradual, peaceful, market-oriented reform on the one hand and violent, youth-driven, Cultural Revolution idealism on the other."[65] Rightly or wrongly, much of the Chinese leadership regards too rapid a political opening as opening the door to fractionalization, neowarlordism, class or ethnic conflict, or economic crisis. As an alternative, the Market-Leninist or soft authoritarian model, assumes a greater attraction—at least for now.[66] It is probable, therefore, that the emerging civic society in China will undergo the same torturous process, ultimately evolving into a more open political system. In the medium term, the post-Deng era will occur and a shift to a new, perhaps initially collective, leadership is likely.

Conclusion

China is clearly on the move. Although in the mid-1990s it cannot be regarded as a full-fledged tiger, it is indeed a cub, stretching its claws and learning how to stalk other capitalist prey. The path forward is going to be difficult and the pace of such things as legal and financial reforms can be expected to be slow, but hardly at a low level of growth. In many respects, China is demonstrating many of the tendencies that made the earlier tigers successful—flexibility and pragmatism in dealing with the international marketplace (especially with exports); an increasing openness to outside investment and managerial expertise; and an effort to improve the standard of living of the vast majority of the citizens at the unspecified cost of political freedoms. In its political development, it clearly aspires to soft authoritarianism. At the same time, it should be underscored that China's development is different from Singapore's, Hong Kong's, South Korea's, and Taiwan's. It is a huge country, with the world's largest population, and what economic advances it makes have a direct political and military weight in regional and global affairs.

Notes

1. China was surpassed in real GDP growth in the 1980–1992 period by Botswana at 10.1 percent, and South Korea at 9.4 percent. World Bank, *World Development Report 1994* (New York: Oxford University Press, 1994), 164–65.

2. Gerald Segal, *The Fate of Hong Kong: The Coming of 1997 and What Lies Beyond* (New York: St. Martin's Press, 1993), vii.
3. Sheryl WuDunn, "As China's Economy Thrives, The Public Sector Flounders," *The New York Times* (16 December 1991): A5.
4. Joyce Barnathan, et al., "China: The Emerging Powerhouse of the 21st Century," *Businessweek* (17 May 1993): 55.
5. Tony Walker, "Investment in China Set To Top $100 bn," *The Financial Times* (8 November 1993): 3.
6. Henny Sender, "Pin-Striped Pioneers: For Accountants, China Is the New Frontier," *Far Eastern Economic Review* (12 November 1992): 59.
7. Craig S. Smith, "Shougang Corp.'s Forays Into Hong Kong Show International Ambitions of China's Young Elite," *The Asian Wall Street Journal Weekly* (week of 7 June 1993): 1.
8. *Businessweek* (17 May 1993): 58.
9. Ibid.
10. Richard Hornik, "Bursting China's Bubble," *Foreign Affairs* (May/June 1994): 4.
11. Ibid.: 26.
12. Salisbury, *The New Emperors,* 145.
13. Ibid., 166.
14. George Rosen, *Contrasting Styles of Industrial Reform: China and India in the 1980s* (Chicago: University of Chicago Press, 1992), 14.
15. Anne Thurston, "The Dragon Stirs," *Wilson Quarterly* (Spring 1993): 12.
16. As Thurston noted: "When Deng assumed power two years later [1978], he set out to implement an old-fashioned, dynastic-style restoration," in "The Dragon Stirs," 12.
17. World Bank, *World Development Report 1986* (New York: Oxford University Press, 1986), 104.
18. Thurston, "The Dragon Stirs," 12.
19. Ibid.
20. Paul Theroux, "Going to See the Dragon," *Harper's Magazine* (October 1993): 41.
21. "China's Remotest Regions Slip Deeper Into Poverty," *The Christian Science Monitor* (22 July 1992): 10–11.
22. Thurston, 14.
23. World Bank, *World Development Report 1991* (New York: Oxford University Press, 1991), 208.
24. Thurston, "The Dragon Stirs," 15.
25. Ibid.
26. William H. Overholt, *The Rise of China: How Economic Reform is Creating a New Superpower* (New York: W. W. Norton & Company, 1993), 31.
27. Walker, "Investment in China Set to Tap $100 bn.," 3.
28. Segal, *The Fate of Hong Kong,* 156; and "Taiwanese Investment in China: Reverse Takeover," *The Economist* (5 December 1992): 66.
29. "Taiwanese Investment in China: Reverse Takeover," 66.
30. Ibid., 57.
31. Lynne Curry, "100 Million Lines by Year 2000," *Financial Times International Telecommunications Survey* (18 October 1993): x.
32. "Telecommunications Survey," *The Economist* (23 October 1993): 15.
33. Gao Jin'an, "China to Launch 11 US-built Satellites," *China Daily* (Beijing) (1 March 1994): 1.

34. Tony Walker and Dierdre Nickerson, "China's Regulators Face An Unenviable Task," *The Financial Times* (25 March 1993).
35. Robert Steiner, "Alongside China's Official Exchanges, Booming Illegal Stock Market Operates," *The Wall Street Journal* (22 June 1993): A11.
36. Allerton G. Smith and Scott B. MacDonald, "China's Banking Evolution," *CS First Boston Fixed Income Research Report* (17 December 1993): 3.
37. Peter McGill, "Shanghai Bids to Become A Top Financial Centre," *Euromoney* (April 1993): 57.
38. Ibid.
39. Ibid.
40. Ibid.
41. Ibid., 62; and "Shanghai Posts 13.6% Growth in First Half; Inflation a Problem," *Bloomberg* (18 July 1994).
42. Ibid.
43. Kathy Chen, "In Cradle of Peasant Revolution, China's Farmers Grow Increasingly Frustrated With Government," *The Asian Wall Street Journal* (7 June 1993): 4.
44. Ibid.
45. For information concerning official corruption see Alan Liu, "The Politics of Corruption in the PRC," *American Political Science Review 77,* 3 (September 1983): 602–23; Wojteu Zanfanolli, "A Brief Outline of China's Second Economy," *Asian Survey 25,* 7 (July 1985): 715–37; Lowell Dittmer, *China Under Reform* (Boulder, Colo.: Westview Press, 1994), 43–45.
46. As quoted in Richard Baum, *Burying Mao: Chinese Politics in the Age of Deng Xiaoping* (Princeton, N.J.: Princeton University Press, 1994), 379.
47. Brett D. Fromson, "Veteran Investor Casts Global Net," *The Washington Post* (11 July 1993): H10.
48. Marcus W. Brauchli, "Chaotic Change: Beijing's Grip Weakens as Free Enterprise Turns Into Free-for-All," *The Wall Street Journal* (26 August 1993): A5.
49. Baum, *Burying Mao,* 379–80.
50. Ibid., 380.
51. "China Speeds on to Market," *The Economist* (20 November 1993): 36.
52. Overholt, *The Rise of China,* 49.
53. Julia Leung, "China Official Tries to Allay Tax Plan Fears," *The Wall Street Journal* (22 October 1993): A11.
54. See Tsang Shu-ki and Cheng Yuk-shing, "China's Tax Reforms of 1994: Breakthrough or Compromise?" *Asian Survey* (September 1994): 769–88.
55. Paul Schulte, "China: Re-Capitalizing the Economy," *CS First Boston Equity Research* (June 1994): 5.
56. Schulte, 5.
57. Quoted from Lena H. Sun, "China's New Ideology: Make Money, Not Marxism," *The Washington Post* (27 July 1993): A5. Some of the same concerns are raised by Orville Schell, *Mandate of Heaven: A New Generation of Entrepreneurs, Dissidents, Bohemians, and Technocrats Lays Claim to China' Future* (New York: Simon and Schuster, 1994), 435–36.
58. Simon Winchester, *Pacific Nightmare: A Third World War in the Far East* (London: Pan Books, 1992), 90–91.
59. Helen F. Siu, "Cultural Identity and the Politics of Difference in South China," *Daedalus* (Spring 1993): 20.

60. Ibid.: 21. Other gloom and doom analysts of China include Hornik, "Bursting China's Bubble," *Foreign Affairs* (May/June 1994): 28–42; and Gerald Segal, "China's Changing Shape," *Foreign Affairs* (May/June 1994): 43–58.
61. Robert Lenzner, "Bearish on America," *Forbes* (19 July 1993): 104.
62. Patrick Tyler, "In the Twilight of Deng; China's Rising Stars Battle," *The New York Times* (21 August 1994): 3.
63. Perry Link, "China's 'Core' Problem," *Daedalus* (Spring 1993): 191.
64. Nicholas D. Kristof, "China Sees 'Market-Leninism' as Way to Future," *The New York Times* (6 September 1993): 1. For a more detailed discussion about market Leninism see Nicholas Kristof and Sheryl WuDunn, *China Awakes: The Struggle for the Soul of a Rising Power* (New York: Times Books, 1994).
65. Overholt, *The Rise of China*, 93.
66. Denny Roy, "Singapore, China, and the 'Soft Authoritarian' Challenge," *Asian Survey xxxiv*, 3 (March 1994): 231–42. Roy argues that this view is also accepted in Singapore, which is an ideological ally of China on the question of political development.

References

Barnathan, Joyce, Pete Engardio, Lynne Curry, and Bruce Einhorn. "China: The Emerging Powerhouse of the 21st Century." *Businessweek* (17 May 1993).

Brauchli, Marcus W. "Chaotic Change: Beijing's Grip Weakens as Free Enterprise Turns Into Free-for-All." *The Wall Street Journal* (26 August 1993).

Businessweek (17 May 1993).

Chen, Kathy. "In Cradle of Peasant Revolution, China's Farmers Grow Increasingly Frustrated With Government." *The Asian Wall Street Journal* (7 June 1993).

"China Speeds on to Market." *The Economist* (20 November 1993).

"China's Remotest Regions Slip Deeper Into Poverty." *The Christian Science Monitor* (22 July 1992).

Curry, Lynne. "100 Million Lines by Year 2000." *Financial Times International Telecommunications Survey* (18 October 1993).

Fromson, Brett D. "Veteran Investor Casts Global Net." *The Washington Post* (11 July 1993).

Hornik, Richard. "Bursting China's Bubble." *Foreign Affairs* (May/June 1994).

International Monetary Fund. *World Economic Outlook, May 1993.* Washington, D.C.: International Monetary Fund, May 1993.

Jin'an, Gao. "China to Launch 11 US-built Satellites." *China Daily (Beijing)* (1 March 1994): 1.

Kristof, Nicholas D. "China Sees 'Market-Leninism' as Way to Future." *The New York Times* (6 September 1993).

Lenzner, Robert. "Bearish on America." *Forbes* (19 July 1993).

Leung, Julia. "China Official Tries to Allay Tax Plan Fears." *Wall Street Journal* (22 October 1993).

Link, Perry. "China's 'Core' Problem." *Daedalus* (Spring 1993).

McGill, Peter. "Shanghai Bids to Become A Top Financial Centre." *Euromoney* (April 1993).

Overholt, William H. *The Rise of China: How Economic Reform is Creating a New Superpower.* New York: W.W. Norton & Company, 1993.

Rosen, George. *Contrasting Styles of Industrial Reform: China and India in the 1980s.* Chicago: University of Chicago Press, 1992.

Roy, Denny. "Singapore, China, and the 'Soft Authoritarian' Challenge." *Asian Survey xxxiv,* 3 (March 1994).

Salisbury. *The New Emperors.* 145.

Schulte, Paul. "China: Re-Capitalizing the Economy." *CS First Boston Equity Research* (June 1994).

Segal, Gerald. "China's Changing Shape." *Foreign Affairs* (May/June 1994): 43–58.

______. *The Fate of Hong Kong: The Coming of 1997 and What Lies Beyond.* New York: St. Martin's Press, 1993.

Sender, Henny. "Pin-Striped Pioneers: For Accountants, China Is the New Frontier." *Far Eastern Economic Review* (12 November 1992).

"Shanghai Posts 13.6% Growth in First Half; Inflation a Problem." *Bloomberg* (18 July 1994).

Siu, Helen F. "Cultural Identity and the Politics of Difference in South China." *Daedalus* (Spring 1993).

Smith, Allerton G. and Scott B. MacDonald. "China's Banking Evolution." *CS First Boston Fixed Income Research Report* (17 December 1993).

Smith, Craig S. "Shougang Corp.'s Forays Into Hong Kong Show International Ambitions of China's Young Elite." *The Asian Wall Street Journal Weekly* (7 June 1993).

Steiner, Robert. "Alongside China's Official Exchanges, Booming Illegal Stock Market Operates." *The Wall Street Journal* (22 June 1993).

Sun, Lena H. "China's New Ideology: Make Money, Not Marxism." *The Washington Post* (27 July 1993).

"Taiwanese Investment in China: Reverse Takeover." *The Economist* (5 December 1992).

Theroux, Paul. "Going to See the Dragon." *Harper's Magazine* (October 1993).

Thurston, Anne. "The Dragon Stirs." *Wilson Quarterly* (Spring 1993).

Tyler, Patrick. "In the Twilight of Deng; China's Rising Stars Battle." *The New York Times* (21 August 1994).

Walker, Tony. "Investment in China Set To Top $100 bn." *The Financial Times* (8 November 1993).

Walker, Tony and Dierdre Nickerson. "China's Regulators Face An Unenviable Task." *The Financial Times* (25 March 1993).

Winchester, Simon. *Pacific Nightmare: A Third World War in the Far East.* London: Pan Books, 1992.

World Bank. *World Development Report 1986.* New York: Oxford University Press, 1986.

______. *World Development Report 1991.* New York: Oxford University Press, 1991.

______. *World Development Report 1993.* New York: Oxford University Press, 1993.

WuDunn, Sheryl. "As China's Economy Thrives, The Public Sector Flounders." *The New York Times* (16 December 1991).

6

Out of the Communist Brush: Slovenia, Vietnam, and Kazakhstan

At first glance, Slovenia, Vietnam, and Kazakhstan have little in common. They share no cultural traditions and are thousands of miles apart. Even their historical evolutions are radically different. While Vietnam has a well-established sense of national identity dating back several centuries, Kazakhstan and Slovenia are recent constructs as nation-states. However, these three countries share a Communist past that is being shed. Slovenia and Vietnam, followed at a distance by Kazakhstan, are leaving the Communist brush—that well-known jungle of obsessive red tape, bureaucratic controls, and caninelike devotion to the gods of central planning. As tigers-to-be they are bound for the market-abundant fields of international capitalism. Slovenia is far out in front in this respect, but Vietnam is already being promoted as the Asia/Pacific region's next tiger. Kazakhstan's progress is more tentative, but can be expected to gain momentum by the end of the 1990s.

While there are no guarantees that the three countries covered in this chapter will be new tigers by the end of the 1990s, it is likely that their advances will be considerably better than most nations in the developing world. Slovenia, for example, stands an excellent chance of joining the European Union early next century, while Vietnam's integration with the rest of the dynamic Asia/Pacific region will advance at a rapid pace. Kazakhstan offers a case where it will benefit from the relative risk levels of investment in the East European region—it already possesses an industrial base, is comparatively stable on the political front, and has a largely literate population. Meanwhile, most of Central Asia faces a difficult future, where the lesser endowment of resources will weigh heavily for foreign investors also concerned with ethnic unrest and the willingness of radical Muslims in Afghanistan and Iran to meddle in local affairs (as in Tajikstan).

Slovenia: Are There Tigers in the Alps?

When Yugoslavia fragmented in the early 1990s, leaving a legacy of ethnic cleansing, ruined cities, and destroyed economies, Slovenia was able to make a relatively clean break. After a brief ten-day war in June, 1991, it suffered only minor damage and gained its independence without the trauma that soon overcame Croatia and Bosnia-Hercegovina. While economic sanctions caused hyperinflation and basic consumer goods shortages in Serbia-Montenegro and fighting continued in Bosnia and Croatia, Slovenia began the process of adjusting to a post-Yugoslavian economy and a democratic political system.

Slovenia is approximately half the size of Switzerland, covering 20,251 square kilometers, and has a population of a little over two million. In sharp contrast to most of the other members of the former Yugoslav federation, the vast majority of the population (91 percent) are Slovenes, with small sprinklings of Serbs, Croats, Italians, and Hungarians. The population is predominantly Roman Catholic and the literacy rate is 100 percent. One of the major problems facing Slovenia is that it is frequently confused with Slovakia, further to the north. Despite confusion and ignorance about Slovenia, it is emerging as one of the economic success stories in the 1990s.

The idea of Slovenian nationalism is relatively new in the context of other European peoples. Around 1274, what was eventually to become Slovenia was made part of the Hapsburg Crownlands, an event that was to leave a strong Germanic imprint on the country in terms of cultural outlook and center of political gravity. As Mark Thompson commented: "The Slovenian temperament is famously Austrian. This means phlegmatic, orderly, hard-working, dull, piously Catholic, reticent, stingy."[1] Despite the Austrian influence (and whatever one may think of Thompson's view of Slovenians), Slovenia did evolve a separate cultural identity that reflected its Slavic roots, with the first books printed in Slovenian in 1550. German domination lasted until 1918 when the Austro-Hungarian Empire collapsed and Slovenia became part of the new kingdom of Yugoslavia. Throughout its Yugoslav tenure, Slovenia (along with Croatia) was the strongest state economy, contributing to the development of the federation's poorer regions. When Slovenia opted for independence, cuts in federal transfers saved the Slovenian budget about 30 percent of its traditional current expenditure.[2]

Since independence Slovenia has made considerable strides in the marketization of its economy and is clearly a tiger candidate. Despite the loss of trading opportunities and factories in other parts of the old Yugoslav federation (such as in Bihac and Tuzla), Slovenia emerged with its economy largely intact and without war damage. Moreover, Slovenian companies have been able to reorient trade away from Serbia-Montenegro and Croatia (complicated by tariffs imposed by that country) toward the European Union and the European Free Trade Association countries (Iceland, Norway, and Switzerland). At the same time, the government has presided over a number of reform measures that have resulted in a nearly convertible, independent currency, the Slovene tolar, supported by strong foreign exchange reserves ($2.5 billion at year-end 1994); a small stock exchange, the Borza, has been established in Ljubljana; and a bank reform program has been implemented. Although the privatization process has been slow in getting off the ground, it became law in December, 1992 and a large number of companies were sold for approximately $6 billion in the 1993–1994 period. Slovenia has also joined the World Bank, International Monetary Fund, and the European Bank for Reconstruction and Development, while attracting increasing investment from Austrian, French, German, and Italian companies.

Slovenia's adjustment process has not been without problems. The economy contracted sharply in 1991 and 1992, largely caused by the collapse of trade with the former Yugoslavia. In 1993 it contracted a further 2 percent. This has also meant that many manufacturing companies have been forced to downsize their workforces. Additionally, the ongoing conflicts further south have hurt Slovenia's tourism—a critical sector. The combination of these factors and the slowdown in European economies in general has translated into high unemployment (11.6 percent at year end 1992) and a decline in the country's per capita income. Per capita income GDP fell from $8,658 in 1990 to $6,052 in 1992.[3] Concerns about high unemployment have also slowed the privatization process. It was only in 1994 that the economy began to turn around, growing 5 percent, with unemployment falling around 9 percent by year-end. By 1995, Sovenia's prospects appeared relatively positive—economic growth was forecast at around 5 percent and inflation was set to fall from 22.9 percent in 1993 to 13 percent for 1995.[4]

Tiger Factors

Does Slovenia have what it takes to be a new tiger? The short answer is yes. Despite the country's small size and lack of natural resources such as hydrocarbons or coal, the Slovenes have a tradition of maximizing their food and energy resources. The Krsko nuclear plant supplies much of the country's electricity needs, while its transport and communications infrastructures are well developed.[5] Although local agriculture does not meet all local needs, the Slovenes have been active in using technology to augment production. The country's Biotechnology Center was established in the late 1980s to coordinate research efforts in biotechnology and agriculture. Particular attention has been given to the improvement of the technology of field crops, plant protection, research into the improvement of agronomic relevant properties of horticultural plants, and forage plants in mountain regions.[6]

Slovenia has a number of other critical tiger advantages. Its political leadership, which is democratically elected, shares a consensus on the country's development goals—marketization and eventual membership in the European Union. An example of this consensus was evident in February, 1993 when the government of Liberal Democratic Prime Minister Janez Drnovsek was able to work with the country's labor leaders and former communists in his ruling coalition to pass a wage pact. Although the agreement will reduce real wages, it was widely agreed that such an action was necessary to reduce inflation and make Slovenian products more competitive.

Slovenia's other tiger advantages include institution building that is supportive of the marketization process. As Anthony Robinson noted in 1993, "the coalition government that emerged from last December's elections [1992] to the national parliament and parallel presidential elections, is determined to use its four-year mandate to complete the transformation of the country into a fully-fledged, market-oriented, multi-party parliamentary democracy."[7] While this observation reinforces the idea of leadership's consensual view, it also reflects that the underpinning institutions involved in the marketization process are developing along a responsive and responsible path. Political instability, such as coups and terrorist campaigns, is not likely to occur in Slovenia and, hence, should not disrupt the country's economic advances. In fact, the civilian authorities presided over an extensive restructuring of the armed forces in March and April, 1994 that included the sacking of the controversial Defense Minister Janez Jansa.[8] A key reason for this is that much of the population regards the country's political institutions with a high degree of legitimacy, a condition reinforced by the holding of local governments in December 1994. Moreover, the rule of law is largely respected, providing a strong foundation for the legal right of economic changes.

Another advantage Slovenia has is ethnic uniformity. Slovenia's population is highly homogeneous and does not have problems with ethnic tensions. This clearly distinguished it from the rest of the former Yugoslavia, the former Soviet Union, and developing countries such as India and Peru.

A strong plus for Slovenia is the high tech factor. As one observer noted in 1993: "The nation's industrial and financial infrastructure is already more westernized than in other east European nations. Its stock markets are well developed, and its telephone system works."[9] Slovenia possesses a well-educated population, much of which has been educated in the West and is plugged into the fast economies in the West. This has meant that Slovenia's technological infrastructure, while not cutting edge, has been and still is competitive. Moreover, the large number of joint ventures with Western firms provide an ongoing link to those economies with the most advanced technology, not to mention sales contracts. Foreign companies, such as Renault and Siemens, have already invested in Slovenia. The motor parts, furniture, high-end textiles, and pharmaceutical sectors in particular are making use of the small Alpine country's relatively advanced industrial infrastructure.

An example of Slovenia's competitiveness in the high-tech range is the Gorenje company, a major appliance maker. Gorenje has tapped the country's well-educated population and emerged as a significant force in the European appliance market because of its tight quality control, attention to research and development, and aggressiveness in searching out new markets in the Middle East and South America. While it sells stoves, refrigerators, cookers, and washing machines under its own name, it also produces many of the same goods under the foreign labels of Bosch, General Electric, Siemens, and Electrolux.

Gorenje is not an aberration for Slovenia. The country's economy is oriented toward exports, the bulk of which are manufactured goods, most of which are from the metal-processing industry, the electrical and electronics sector, motor vehicles and other transportation equipment, chemical, textile, and wood-processing industries. On a year-to-year basis these products constitute roughly 70 percent of Slovene exports.[10]

The major concerns for the advancement of the Slovene economy are generated beyond the country's frontiers. These include the potential for the spread of hostilities between Serbia and Croatia or a new conflict between Serbia and Hungary over the treatment of minorities. Additionally, the European Union remains important to Slovenia's future plans. If the EU takes a stand against new memberships after Austria, Finland, and Sweden (who became members on 1 January 1995), this could hurt the small Alpine country's prospects.

Slovenia has found the tiger track in the 1990s. Its departure from the Yugoslav federation, movement toward a Western multiparty democracy and market economy, and membership in international bodies reflect a process of change that is significant. While much of Central and Eastern Europe grapples with throwing off the traditions and attitudes of the old Communist order, Slovenia is racing along, seeking to be even more plugged into the fast economies. Its prospects are very strong in the tiger category.

Vietnam: Finding Its Stripes

The transformation of the Vietnamese economy in the 1990s has been long awaited by foreign investors and badly needed for the country's population. Considering the significant economic advances achieved by Vietnam's neighbors, such as Singapore, Thailand, and Malaysia, the question is—What took so long? The answer is that the Vietnamese "revolution" that succeeded in reunification of a victorious Communist North and a capitalist, yet U.S.-dependent South, precluded the emergence of a tiger in Indochina for two decades after the fighting stopped in 1974. Instead, Vietnam receded into the Communist brush, where Marxist-Leninist ideology triumphed over economic realities. Proclamations about the advances in the creation of socialist society were matched by chronic balance of payments deficits, low productivity in both agriculture and industry, and ongoing tensions between an entrepreneurial south and a doctrinaire north. By the late 1980s it became evident to Hanoi's aging leadership that party pragmatists were correct in advocating the use of market mechanisms, such as profits and incentives, to stimulate growth. Otherwise, Vietnam ran the risk of falling further behind most of the Asia/Pacific region in terms of economic productivity and most basic human indicators.

A Brief Background Note

When the last U.S. helicopter left the skies above Saigon in 1974, the Communist leadership in the North had accomplished their goal of reunification. Surprisingly it was achieved by military means and in a short period of time, not as expected over a long period and by negotiation. Although the consolidation of power was conducted without a massive bloodletting (as in the case of neighboring Cambodia), the challenges that lay ahead were massive. Yet, over thirty years of fighting (beginning against the Japanese in the 1940s) left Vietnam unified and with a powerful and victorious military and a leadership who finally tasted the success of victory over seemingly impossible odds. Policy planners in Hanoi, after some initial hesitations, moved to advance central planning in the South. This action sought to unify the country's economic development.

That development was complicated from the very beginning. The lengthy separation of Vietnam's two regions created a major complication in the implementation of a socialist economy. The North, traditionally overpopulated with its productive base centered in the Red River Delta, had implemented central planning in the 1950s and maintained this approach into the unification period. Private enterprise was allowed, but only within tightly controlled circumstances. Despite a steady stream of propaganda about the wonders of socialism, the simple fact was that the Northern economy remained structurally weak, internationally isolated (because of U.S. hostility), and burdened by the costs of the war in the South. Additionally, it was damaged by U.S. aerial bombardment, and dependent on economic assistance from the Soviet Union and, to a lesser extent, the People's Republic of China.

The South, in sharp contrast, had several decades of capitalist experimentation (albeit with cronyism) by the time of reunification. Although the South's capitalist period was supported by a massive infusion of

U.S. economic and military assistance, it nonetheless reinforced the existence of a local entrepreneurial class, especially within the overseas Chinese community. Moreover, the long period of conflict in the South left a highly urbanized society as many peasants fled the violence in the countryside. Unlike the North, consumer goods, awareness, and links to the larger world, and an understanding (in the broadest sense) of capitalism had become part of the South's perceptual outlook. The differences between North and South were furthered by the decimation of Communist cadres in the latter during the war and the larger number of northerners placed in commanding positions over their southern countrymen in the aftermath of the conflict.

Hanoi's primary concern in the mid-1970s was how to reactivate economic growth through socialism. Operationally this meant the dismantling of capitalism in the South and returning that region to its traditional role as the nation's breadbasket. In 1977 and 1978 a tough campaign was launched to close down all small private businesses in the South, while the 1976 Fourth Party Congress' Second Five-Year Plan unrealistically outlined production goals beyond the economy's means. Simply stated, the newly unified economy was centrally planned with impossible targets. As Frederick Z. Brown noted: "Heavy industry was stressed, farms and light industry in the South were collectivized, the need to give farmers and workers incentives was disregarded. And, in practice, it all worked about as might be expected, which is to say, not at all."[11]

As no large increase in foreign assistance was forthcoming—China in fact reduced the amount—Vietnam was forced into a position of greater self-sufficiency. This situation was complicated by ongoing hostility on the part of the United States, which refused to pay reparations agreed upon at the Paris Peace Talks due to the North's military conquest and questionable accounting of soldiers missing in action and prisoners of war. Furthermore, Vietnam's relations with its immediate neighbors, China and Cambodia, became increasingly tense in the 1970s. This was significant for two reasons: the deterioration in relations with Beijing and Phnom Penh meant ongoing high military costs for Hanoi and the failure of regional trade to develop reduced prospects for the Vietnamese economy's recovery. This situation worsened dramatically when Vietnamese forces invaded Cambodia in 1979, ousting, but failing to destroy the Khmer Rouge. A brief, but bloody border conflict with China in the same year maintained pressure on the northern border. These factors made Vietnam rely more heavily on its sole major international ally, the Soviet Union.

In the late 1980s, Vietnam's leadership underwent a number of changes. First and foremost, a number of the old Communist vanguard died or retired from active politics. Secondly, a younger and more pragmatic generation of leadership regarded past economic policies largely as failures. This was especially evident when Vietnam's experience was compared with its neighbors, Thailand and Malaysia, and most recently China. Thirdly, the costly military occupation of Cambodia was regarded as a severe burden on the Vietnamese economy—it was too much a strain if improvements at home were ever to be made.

Making the Shift to Tiger Cub

Vietnam's shift from communist elephant to capitalist tiger cub commenced in 1986 when the national leadership finally adopted a more pragmatic line called *doi moi,* or renovation. Careful to maintain political control through a single party state—similar to some extent to China's market-Leninism—Vietnam launched a comprehensive economic adjustment program that included a loosening of controls on the private sector, opening the economy to foreign investment, and a degree of liberalization in the trade regime. Although Vietnam's turn to the market was nowhere as dramatic as China's or Central and Eastern Europe's, it nonetheless was a significant change.

The desire to accelerate the pace of change led the government to implement a new adjustment program in 1989. At the time, the experimentation with the market had not yet produced the hoped-for results and structural imbalances required stabilization. Rising inflation was a particular problem, reaching over 300 percent in 1988. Another important concern was that large-scale foreign investment had not poured into the country. Consequently, the 1989 program established the State Committee for Cooperation and Investment (SCCI), which was given the mission to simplify the procedures for foreign investment. Other provisions in 1989 included new measures to channel investment funds into strategic areas such as high-tech industry and infrastructure.[12]

An important factor in Vietnam's success is its turning to foreign investment as an instrument to help stimulate local growth. Although this should not be overstated due to the U.S. embargo, Vietnam demonstrated that it was aware that foreign capital was badly needed to revitalize the economy. A key element in bringing in a modest amount of foreign investment was the 1989 investment code, which was subsequently updated and ratified in April, 1992. This gave Vietnam one of Asia's most liberal investment codes, providing an express guarantee (now enshrined in the amended constitution passed in April, 1992) against expropriation or nationalization, guaranteeing the right to repatriate capital and profits (subject to 10 percent withholding tax in most circumstances), and imposing no minimum or maximum investment limits and required equity.[13]

Vietnam's transformation from a closed door to a welcome mat for foreign investment is evident in a number of ways. For example, its energy sector has long needed outside expertise and capital for development. In December, 1992 a significant step was taken to accomplish this, when Australia's Broken Hill Propriety and the Malaysian state oil firm Petronas were granted the authority to develop the promising Dai Hung (Big Bear) oil field off the southern coast.[14] The field is estimated to hold 700–800 million barrels of crude oil, which could help meet Vietnam's energy needs or supplement the country's exports.

Vietnam's stabilization and adjustment program is an ongoing success. From 1989 to 1992, real GDP grew at an average annual rate of 6–8 percent, the government's budget did not run any significant deficits, and inflation fell to 17.5 percent by the end of the period.[15] In 1993, the economy grew by 8 percent and inflation fell to 8.0 percent and growth remained strong in 1994 at roughly 9 percent. These gains were made despite ongoing trade restrictions on the part of the U.S. and the Soviet Union's abrupt withdrawal in 1990 of almost all economic support. Vietnam's economy, in fact, emerged through this period in a stronger position than before, with no balance of payments crisis, and poised to gain its tiger stripes.

Vietnam's success in the 1989–1992 period is largely attributed to its ability to maximize its food and energy resources. Vietnam's economy has traditionally been dominated by agriculture, which accounted for 40 percent of GDP through the 1980s. As already mentioned, one of the primary concerns of the nation's leadership in the aftermath of reunification was how to reactivate the South's productive capacities in this sector. Significantly, when the reform process commenced in the mid- and late 1980s, agricultural reforms led the way. They included an easing of government controls and a reduction in the state's overall control in the sector.

The adjustment program in 1989 further stimulated commercial agricultural production. In particular, families were given the right to work their own land and sell their produce at market prices (which incidently helped push consumer prices upwards). The result of reforms in agriculture was as one source notes: "Vietnam transformed itself over a period of two years from a rice importer to the world's second or third largest rice exporter."[16] At the same time, state sector ownership was only 2 percent of agricultural production by 1992.[17]

The combination of rising agricultural exports and Vietnam's tapping of its hydrocarbon wealth helped stimulate growth. Crude oil production, assisted by European and Japanese companies, earned foreign exchange and kept the budget well stocked with revenues. Moreover, Vietnam's domestic energy needs were easily met. This was critical to the expansion of a growing private sector, producing light consumer industrial goods and services in an area of the economy where the state sector was, and remains small.

Vietnam's leadership, though concerned about the "polluting" influence of capitalism, had reached a general consensus about the nation's direction. Like China, Vietnam's Communist party is facing the difficult balancing act of political control with economic freedom. In a sense, Vietnam has opted to follow the market-Leninist approach to development. This approach has allowed it to survive and overcome problems with inflation, endemic corruption, the confusing mix of socialist rhetoric and capitalist economic policies, and a U.S. trade and investment embargo (which is gradually eroding in the mid-1990s). Although problems continue to plague the Vietnamese economy, the Asian country increasingly offers a foreign investment site for companies seeking skilled low-cost labor, a welcoming foreign investment code, and eventually, a sizeable market.

Problem Areas

While Vietnam has maximized its food and energy resources and its leadership appears to have adopted a consensual approach, Vietnam does have serious shortcomings that must be addressed for it to move beyond being a tiger cub to an adult tiger. First and foremost, its technological and industrial infrastructure is weak. The Vietnamese government estimates it needs $50 billion by the year 2000 from international and domestic sources to continue the pace of economic reform—most of which is earmarked for infrastructure upgrades.[18] Although the situation is not bleak and decidedly better than many developing countries (i.e., most of Sub-Saharan Africa, Laos, and Cambodia), a considerable effort is necessary for Vietnam to close the gap with its neighbors, Malaysia, Thailand, and Singapore. For example, many of the country's 8,000–9,000 state-owned enterprises operate only antiquated Russian or Chinese equipment.[19] Also, much of the South, where the most dynamic growth is occurring, lacks adequate electricity supply. Other problems include a national transportation system that is still recovering from the Vietnam war and a near nonexistence of telecommunications systems.[20]

Vietnam does have a national Center for Scientific Research and many of its "best and brightest" students were sent to learn sciences in friendly countries, such as the Soviet Union. In 1992, about 7,000 Vietnamese had earned scientific doctorates from foreign universities and were busy applying their expertise to the local economy.[21] While much of Vietnamese research is oriented to reducing the country's import bill by copying foreign technologies, Vietnam's tech infrastructure remains low-tech, not yet capable of helping plug it into more fast track economies. Additionally, even before Vietnam can opt to pursue high-tech activities, low-tech demands remain central. This includes improving roads, upgrading railroads and ports, and installing power lines.

A second problem area is the lack of proper capitalistic institutions. Institution building with a concomitant legal infrastructure is critical. Foreign investors have put money into Vietnam, but the volume is far lower than in many other developing countries. The lack of a comprehensive legal framework does not inspire confidence, despite the pledges in the constitution. As John Brinsden, Standard Chartered's representative in Ho Chi Minh City, commented in 1992 of foreign investor concerns, "They're worried that if there is a dispute, they could lose everything without recourse to the courts."[22] The Vietnamese government has sought to improve this situation, but concerns remain.

Foreign investment in Vietnam is led by Taiwan, which had seventy-seven projects and a cumulative capital investment of $1.1 billion as of April, 1993.[23] Hong Kong and France are second and third with cumulative investments of $807 million and $488 million, respectively. For

Vietnam to make it to the takeoff stage, it is estimated that the price tag will be $2–3 billion annually of foreign investment over the decade to the year 2000.[24]

The ending of the U.S. economic embargo on Vietnam in February, 1994 is likely to help, in the medium term, the Southeast Asian country gain further access to foreign direct investment. Although U.S. companies are not rushing en masse to the newly opened market, many are carefully sizing up where opportunities may exist. Part of the problem is, as the *Far Eastern Economic Review*'s Murray Hiebert noted, those companies that rush in "will find themselves at the back of a slow-moving queue."[25] That stated, Coca-Cola, Chrysler, and Pepsi either have returned or are exploring setting up operations in Vietnam.

Vietnam also lacks strong financial institutions. Its banking system is inadequate to supply capital to a growing private sector and its capital markets are nascent, a clear disincentive to foreign investors. One reason for the dismal state of Vietnam's financial sector is that its banks are weakened by bad debts. As in many communist economies, the banks were forced to assume the nonperforming loans from the inefficient state industrial sector. Efforts to create state-owned commercial banks to service the growing private sector also have proven inadequate. This situation led to the creation of credit cooperatives, most of which were not regulated and eventually collapsed due to elaborate pyramid schemes. Another factor that maintains weak domestic banks is a lack of accurate and reliable data that makes risk assessment and corporate analysis difficult.[26] To help stimulate credit availability, the Vietnamese government has gradually allowed foreign banks to establish operations.

Vietnam's capital markets have been slow to develop. This is because of the lack of an adequate legal framework and reluctance of the government to privatize or close nonviable industrial enterprises due to concerns over unemployment. Having watched the privatization process in Central and Eastern Europe, some of which was done at fire-sale prices, Vietnamese authorities became concerned about social chaos. Indeed, the word "privatization" is not used in official discussions on the matter, with "equitization" used instead. As Hanoi economist Do Duc Dinh commented, "Privatization suggests something not good: throwing workers to the streets, selling socialist assets, turning Vietnam totally capitalist."[27] This attitude has translated into a slow approach to developing capital markets. As of early 1995, Vietnam still lacked a stock exchange, through the idea was being discussed.

In late July, 1994 Vietnam took an encouraging step to establish a capital market. It was announced that foreigners and overseas Vietnamese were not eligible to buy bonds issued by the Vietnamese government. Although much more needs to be done to develop a full-blown capital market, the measure indicated that some momentum clearly exists. The incentive is that the government needs to raise up to $55 billion in investment (by one account) to double its annual per-capita income of $220 by the end of the decade.[28]

Foreign debt is another problem facing Vietnam. The country's total external debt is estimated at $15 billion in free convertible and nonconvertible currencies.[29] This is a substantial burden on Vietnam. The situation is worsened by U.S. hostility to Vietnam at the IMF. However, the Clinton administration eased the U.S. line to allow Vietnam's relations with its multilateral creditors to improve. This was evident in October, 1993 when Japan and France, with U.S. acquiescence, stepped in to eliminate Vietnam's debt arrears to the World Bank and the Asian Development Bank. This was a critical step in Vietnam's normalization of relations with its creditors, which then allowed Hanoi to reach an agreement with the IMF for a $223 million standby loan.

One last major concern for Vietnam's ability to stay on the tiger track is the potential for North-South tensions. Most foreign investment has been attracted to the South and clearly that region maintained stronger economic growth than the North. While this is not an issue of ethnic difference, it is one of potential regional conflict. Both areas have less-than-flattering perceptions of each other that are based on long-standing historical developments, including the recent war of reunification in the post-World War II era.[30] While these perceptions are often superficial, they mask over the fact that the southern economy is making rapid advances and attracting foreign investment and the northern economy remains poorer, less plugged into the global system, and increasingly dependent on southern successes to help rebuild northern infrastructure. As in China, Vietnam's central authorities will have to carefully weigh how much authority they devolve to the provinces.

Vietnam's ability to stay on the tiger track is less certain than Slovenia's. However, a certain momentum is evident that bodes well for its hardworking and literate population. Much depends on how the government manages new challenges that will test its commitment to economic development versus ideological posturing.

Kazakhstan: Will a Purr be Heard from Central Asia?

The collapse of the Soviet Union in 1991 and the birth of new nations out of the old was particularly dramatic in Central Asia, where five Soviet Republics suddenly found themselves dealing with the many challenges of independence. This situation, of course, was complicated by the unexpected speed of events, a lack of preparation for the burdens of sovereignty, and questions about "national" borders in the face of overlapping ethnic communities. Moreover, the economies of the former Soviet Central Asia were characterized by a heavy agricultural orientation and a range of mineral resources, including in some cases oil and natural gas and, in some cases, heavy industry (i.e, defense and steel). These riches, however, were not evenly divided. One of the more fortunate Central Asian republics to emerge is Kazakhstan. While Kazakhstan has many problems like its neighbors—poverty, heavy environmental pollution, and an underdeveloped industrial infrastructure—it does have certain attributes that provide some hope for the future.

Kazakhstan covers 1,059,750 square miles, or about 12 percent of the former Soviet Union, making it about four times the size of Texas. It has a population of 17 million people and some thirty ethnic groups are represented, of which the Kazakhs and Russians account for almost 40 percent and 38 percent of the population, respectively.[31] The third largest group are the Germans, who account for 6 percent of the total population. Although a certain degree of tension exists between the various ethnic groups, there has not been any major clash nor any substantial deterioration in the country's political stability, as in the case of neighboring Tajikstan.

The roots of contemporary Kazakhstan can be traced back to the fifteenth century and were founded on a mixture of Turkic Moslem and Mongol peoples. The Kazakhs were largely nomadic, bordering the Uzbeks and Kyrgyz peoples in the south and east. The Russian expansion into modern Kazakhstan commenced in the sixteenth century, beginning with military conquest and followed by settlers in the vast steppes.

Kazakhstan spent several centuries under Tsarist rule, largely as a producer of agricultural goods, namely cotton. Only in 1917, when Russia was convulsed by the Bolshevik Revolution, did local Kazakhs assert their independence. That interlude, however, was short lived. By 1919 the Bolsheviks returned to take control, suppressing local forces. Initially attached to Russia, Kazakhstan was designated as a separate union republic in 1936. The republic's economic development remained largely agricultural, with some light and heavy industry. Economic affairs, however, were severely disrupted by Stalin's forced collectivization campaign, which was responsible for the death of as much as one-third of the population and most livestock.[32] During the Second World War, Kazakhstan became an important rear area for the Soviet war effort and was the destination point for Germans uprooted from elsewhere in the Soviet Union by a Stalin apprehensive of a potential fifth column.

Kazakhstan enjoyed a degree of recovery in the 1950s under Khruschchev's "Virgin Lands" scheme. This program involved irrigation and cultivation of new lands (usually former desert). It soon increased grain and livestock production. There was a major downside to the Virgin Lands scheme as its methods ultimately resulted in environmental degradation, most notable in the shrinkage of the Aral Sea, and human suffering.

Kazakhstan's economy at independence in 1992 can only be described as "undeveloped."[33] From the 1950s through the 1980s, the agricultural sector was promoted, especially the cultivation and export of cotton. Other foodstuffs were produced and the surplus exported. At the same time, Kazakhstan's industrial sector was heavily oriented to iron production. However, the industry is dominated by inefficient large state-owned companies. Because these state-owned firms have been and continue to be poorly run, they operate at a loss and place a severe strain on the government budget. This was reflected, in part, by Kazakhstan's fiscal situation in 1988, where the budget deficit stood at a high 8 percent of GDP. Although measures have been taken to contain the deficit, the state sector continues to be a drag, making the fiscal deficit 6 percent of GDP in 1992.[34] The end of the Soviet Union also witnessed a massive contraction in Kazakhstan's economy—at 13 percent in 1991, 14 percent in 1992, 12 percent in 1993, and 6 percent in 1994.[35]

Kazakhstan, like much of the former Soviet Union, has a number of problems. The most pressing is environmental damage. This was caused by intensive irrigation techniques, which contributed to the evaporation of the Aral Sea and potentially threatens a severe water shortage. An additional point of concern is that nuclear testing in parts of the country for several decades has further polluted a number of areas, making them unhealthy for either human habitation or agricultural cultivation.

Kazakhstan also faces a number of challenges on its borders. The post-Soviet era has not left a peaceful interlude in Central Asia, but just the opposite. Alma Alta's security concerns emanate from its borders with Uzbekistan, Kirghizia, and Russia as well as sharing the Caspian sea with a civil war-torn Azerbaijan and a militarily rebuilding and potentially aggressive Iran. A particular concern is the Tajik civil conflict, which has pitted the former communists, with Russian military support, against a coalition of Islamic militants and liberals. In 1992–1993 the conflict was brutal, with heavy fighting in the capital and along the Afghan border. Although the former communists were eventually able to emerge in control of the capital, an Islamic government-in-exile in northern Afghanistan promises to maintain tensions in the region.

Does Kazakhstan Have Tiger Ingredients?

Considering the daunting set of challenges, initial inspection of Kazakhstan does not make it a tiger. However, the country does possess a number of critical ingredients that are likely to put it on the right track by the end of the 1990s. These include strong political leadership, a wealth of resources that is attracting badly needed foreign investment, and a relatively well-educated population.

Unlike a number of other former Soviet Republics, Kazakhstan has political stability. President Nursultan Nazarbayev, a Kazakh and former chairman of the Kazakh Cabinet of Ministers (under the Communist regime) became president in an unopposed election in December, 1991. Although a myriad of political parties emerged with independence, real political power remains with Nazarbayev and the Socialist Party of Kazakhstan (the former Communists).

Nazarbayev's presidency is quasi-authoritarian, much in the fashion of Asian newly industrializing economies at earlier stages in their development. This translates into elections and dissent being allowed within clearly delineated limits. Parliamentary elections were held on 7 March 1994; roughly 74 percent of the electorate turned out to vote, casting their ballots overwhelmingly for the ruling party. Despite claims about voting irregularities, Kazakhstan remained politically calm and the government very much in control.[36] The government has indicated that it wants a secular political system, a strong presidency with limited checks, and is supportive of a multiethnic society.[37] While the Kazakh language is now the official state language, Russian has been maintained as the Republic's means of communication, hence providing two languages to be used in the new nation's daily life and safeguarding the linguistic rights of a sizeable and economically significant minority.

Nazarbayev's claims that his somewhat heavy-handed rule is necessary in a time of transition has been partially vindicated by the problems elsewhere in the former Soviet Union. Moreover, local political groups exist that could fragment Kazakhstan's political system. These include the Azat Social-Democrat Party, which advocates an end of Russian immigration into the republic and the Alash and Zhektoqson parties. The latter are Kazakh nationalist parties that have been refused registration and official recognition. Alash stands for decollectivization of agriculture, redrawing borders to include traditional Kazakh cities now outside the republic, repatriation of non-Kazakhs, and Turkic and Muslim solidarity.[38] Although these groups are not strong, they do represent radical forces that could open the door to bigger problems. As Assan Nougmanov, President of the Kazakhstan Association for the Advancement of International Scholarly Projects and Exchanges, noted: "It is vitally important for Nazarbayev to separate the demands for a Kazakh cultural renaissance from the demands by nationalists that unconditional political and economic priority be given to the indigenous nation, as opposed to other ethnic groups. The Slavs' adaptation to life in a Kazakh national state must be orderly and gradual."[39] Considering that balancing act, Nazarbayev's strong leadership has provided a relative degree of political order. The importance of his leadership was noted by Paul B. Henze. "Despite his background as a communist, Nazarbayev comes the closest of any current Central Asian leader to showing the kind of determination and political skill of the man Central Asian leaders could well aspire to emulate: Mustafa Kemal Ataturk, founder of the modern Turkish nation."[40] Equally significant, the Kazahk leader has a vision for the future, which is strongly influenced by Korea's development experience. It is no mistake that Nazarbayez has visited and studies Korean conglomerates Samsung and Lucky Goldstar.

Kazakhstan also possesses a wealth of natural resources, which is helping the country to emerge from its difficult birthing. Its resources include gold, oil, copper, zinc, titanium, magnesium, and chromium, as well as substantial coal and uranium supplies. While Kazakhstan does not have the expertise to tap all these resources, foreign investors do. In one survey examining foreign investment in the post-Communist bloc, foreign investors committed more than $42 billion to more than 1,700 projects in twenty-eight countries in the eighteen months up to March, 1993—of that Kazakhstan was number two, with $9.12 billion pledged or committed.[41] Foreign ventures into Kazakhstan include the involvement of companies, such as British Gas, Agip of Italy, and Chevron. Philip Morris, the U.S. tobacco and food products group, became the first foreign company to acquire a significant portion in a privatized company in Kazakhstan by picking up 49 percent ownership of Almaty Tobacco Kombinat, a cigarette manufacturing operation with 1,700 workers.[42]

While natural resources are not regarded as a tiger factor per se, in the case of Kazakhstan oil revenues could provide enough wealth to allow a less painful transition to a market economy (see table 6.1). In 1993, after an agreement with Chevron, which splits earnings 20/80 in favor of Kazakhstan, the vast Tenghiz oil field in western Kazakhstan was opened. The value of this deal should not be understated. As *Far Eastern Economic Review* journalist Carl Goldstein noted: "With an estimated 6–9 billion barrels of recoverable oil, the field could in time bring Kazakhstan riches far in excess of any Western aid package for its Russian neighbor. Even if the lower estimate of reserves proves correct, the field should yield at least $114 billion worth of oil over its expected 40-year life, based on current prices."[43]

Kazakhstan's educational facilities, necessary for it to purchase and assimilate badly needed high technology, have the potential to provide a foundation for the future. The Central Asian country's literacy rate is 84 percent, which is better than many developing countries, such as India, Peru, and Morocco. Kazakhstan has 8,064 secondary schools, 244 secondary specialized schools, and 55 higher schools, including universities.[44] This has translated into the highest number of specialists with higher education and scientists in Central Asia, which has helped make Kazakhstan more industrialized than its neighbors. At the same time, the educational system is in need of an upgrading in many areas and educational facilities are highly concentrated around cities and towns, creating a gap between a better educated urban population and those living in rural areas.

TABLE 6.1
Kazakhstan's Longer Lasting Oil Wells

	How big are reserves? (billion barrels, end 92)	How long will they last? (Years*)
Selected countries		
Saudi Arabia	257.8	82
Kuwait	94	>100
Russia	48.4	16.7
U.S.	32.1	9.8
Libya	22.8	41.2
Kazahkstan	21.5	>100
Oman	4.5	17.1
Qatar	3.7	21.6
Turkmenistan**	1.5	38.9
Brunei	1.3	20.3

*Reserves at the end of 1992 divided by production in 1992.
**1991 figures.
Sources: Union Bank of Switzerland (Kazakhstan energy ministry, official CIS sources, BP Statistical Review of World Energy).

Kazakhstan does have substantial food supplies, especially in grain production. Grain remains a major export in the postindependence era. In the former Soviet Union, the northern part of Kazakhstan usually produced 28 percent of the U.S.S.R.'s total wheat crop. However, the new republic has difficulty in moving food to the market place. The critical problem areas are transport of goods and bottlenecked distribution networks.[45] Consequently, much needs to be accomplished before Kazakhstan's food resources are maximized. An additional concern is security of water supplies, a potentially difficult issue in the years to come.[46]

Progress on other fronts is mixed. In 1993, a new property law was enacted that allowed for the buying and selling of land leases and use of leases as collateral to borrow money. Tariffs were lowered in 1994, but trade has yet to fully take off. Although privatization of the land remains distant, perhaps within the three- to five-year range, the government has targeted other sectors—housing, trade, services, the wholesale goods sector, and transport—for sale to the private sector. In the early 1990s, transportation underwent a change when "gypsy" taxis started operations. Despite these small beginnings, the task of privatization of significant parts of the economy lags. In 1991, the state sector accounted for close to 90 percent of fixed assets in the economy, around 80 percent of output, and 75 percent of employment, a situation that had not changed considerably by 1994.[47] That stated, the official strategy for 1992–1995 makes privatization and improved supply of consumer goods priorities.

The Gradual Approach to Tigerhood

Kazakhstan's transformation to a market economy is occurring in a gradual fashion. Although economic reform was advocated as early as the 1980s during the perestroika period and it has been acknowledged that there is no alternative to switching to a market, Kazakhstan's economy is still transitional and will be that way for a long time to come. This has meant a large contraction in the economy and high inflation. The fundamental reason is political. As Nougmanov noted, "It was important to Nazarbayev that the crumbling system not bury him under its rubble, and so he did everything within his power to make the collapse gradual and evolutionary, in accordance with the more traditional Kazakh mindset."[48] Too rapid a pace of change could stimulate ethnic differences between Kazakhs and Russians, something that the government has worked hard to avoid. This gradualistic approach, however, is fraught with problems, such as determining when Kazakhstan's currency, the *terge,* will become convertible. This, in turn, raises questions about inter-Central Asian trade as well as the ongoing trade and commercial relationship with the rest of the former Soviet Union.

Kazakhstan's economic reform process will follow a delicate path of balancing foreign investment, promoting exports and the tapping of natural resources with Kazakh aspirations of national identity and the pressing need to modernize entire sectors of the economy. The balancing act requires strong, not necessarily authoritarian, leadership, a more clearcut program for privatization, and promotion of the private sector. The process of transformation is occurring also in terms of attitudes and perceptions. Kazakhstan is a new country with a long history and its citizens need to sort out the past and fill in blank spots during the Soviet period—a difficult process. Despite the daunting challenges facing Kazakhstan, its leadership has not panicked and is committed to moving the country forward at its own pace.

Out of the group of countries to emerge from the communist bush in Central Asia, Kazakhstan probably carries the greatest chance for success as a tiger. It has a large and relatively well-educated population, links to Russia (through its large Russian population) and Germany (another large minority), marketable natural resources, and a higher degree of political stability. While there are many negative factors to consider, the positive factors are likely to become more significant as the decade progresses and foreign investors become dissatisfied with conditions in Russia and other parts of the former Soviet Union. In the early 1990s, foreign investors are waiting for macroeconomic stabilization, new laws (that lock in the marketization process), and stronger market conditions. By the end of the 1990s these factors are likely to be in place.

Conclusion

The group of countries represented in this chapter are likely to be part of the next generation of tigers in the late 1990s. Each is unique in its own history as well as its own approach to economic development, but all are linked by their likelihood in emerging successfully from the Soviet orbit. Like many of the other countries examined in this book Slovenia, Vietnam, and Kazakhstan must wrestle with the problems of rapid economic changes, social displacement, and political concerns linked to the marketization process. These factors hold Slovenia, Vietnam, and Kazakhstan together with China, Argentina, and Morocco. As a group, they have strong leadership (both democratic and nondemocratic), consensus within the national elite, followed a sequencing of economic reforms (Kazakhstan the least), and maximized to some extent their food and energy resources. Within the three countries analyzed in this chapter, Slovenia is clearly the most advanced, followed by Vietnam and Kazakhstan.

Notes

1. Mark Thompson, *A Paper House: The Ending of Yugoslavia* (New York: Pantheon Books, 1992), 18.
2. European Bank for Reconstruction and Development, *Annual Economic Review 1992* (London: European Bank for Reconstruction and Development, 1993), 50.
3. Anthony Robinson and Laura Silber, "Now the Economy Has to Adjust," *Financial Times Survey Slovenia* (30 March 1993): 34.

4. "Drnovsek sees five percent economic growth," *Balkan News International* (16–22 October, 1994): 19.
5. Mark Thompson, "Slovenia—Out of the Maelstrom," *Business Europa* (January/February 1993): 20.
6. The Ministry of Science and Technology of the Republic of Slovenia, *Science in Slovenia: Overview with Highlights* (Ljubljana, Slovenia, 1992), 53.
7. Anthony Robinson, "Republic of Slovenia," *Financial Times Survey* (30 March, 1993): 33.
8. *RFE/RE News Briefs* (11-15 April 1994): 16.
9. Rob Urban and Laura Zelenko, "Europe Today: Balkans War Boosts Slovenia, on Battle's Border," *Bloomberg Newswire* (11 October 1993).
10. Centre for International Cooperation and Development, *Slovenia: New European Partner* (Ljubljana, Slovenia: Centre for International Cooperation and Development, 1992), 25.
11. Frederick Z. Brown, "Vietnam Since the War," *Woodrow Wilson Quarterly* (Winter 1995): 76-77.
12. Sam Y. Kim, "Vietnam's Return to Capitalism: The Next Economic Miracle in Asia?" *Harvard International Review* (Fall 1992): 46.
13. David Frank, "Vietnam: The Great Leap Sideways," *Euromoney* (June 1992): 39.
14. "Vietnam—Oil Field Awarded," *Far Eastern Economic Review* (14 January 1993): 55.
15. Sara Kane, "Transition to a Market Economy: Lessons from the IMF's Experience," *IMF Survey* (26 July 1993): 240.
16. Kane, "Transition to a Market Economy," *IMF Survey*: 240.
17. Frank, "Vietnam," *Euromoney:* 37.
18. Lisa Jane O'Neil, "Vietnam Commercial Debt Investors Waiting for Pay-Off," Bloomberg (24 February 1995).
19. Murray Hiebert, "First Things First," *Far Eastern Economic Review* (11 July 1993): 64.
20. Kim, "Vietnam's Return to Capitalism," *Harvard International Review:* 47.
21. "Vietnam's Struggle for Good Science on the Cheap," *The Economist* (18 April 1992): 87-88.
22. Quoted from Frank, "Vietnam: The Great Leap Sideways," *Euromoney:* 39.
23. Nick Freeman, "Birth of a Tiger Cub," *The Banker* (June 1993): 25.
24. Frank, "Vietnam," 37.
25. Murray Hiebert, "Do It Yourself," *Far Eastern Economic Review* (31 March 1994): 62.
26. "Foreigners Set Up Shop," *The Banker* (January 1993): 55.
27. Quoted from Barry Wain, "Vietnam, Put Off by Privatization Style of Eastern Europe, Takes a Slower Road," *The Asian Wall Street Journal Weekly* (24 August 1992): 2.
28. "Vietnam's Government Makes Bonds Available to Foreign Investors," *Asian Wall Street Journal* (2 August 1994): 5.
29. Kim, "Vietnam's Return to Capitalism," 55.
30. "Vietnam: Still Divided," *The Economist* (4 April 1992): 40.
31. See Jim Nichol, "Kazakhstan: Basic Facts," *CRS Report for Congress* (26 June 1992). Also see Catherine Poujol, with the assistance of Pierre Gentelle, editors, *Asie Centrale: Aux Confines des Empires, Réveil et Tumulte* (Paris: Editions Autrement, 1992).

32. Jim Nichol, "Kazakhstan: Basic Facts," *CRS Report for Congress* (26 June 1992): 1.
33. The Barclays ABECOR Country Report, *Central Asian Republics* (June 1993): 1, regarded the economy of Kazakhstan as "undeveloped."
34. "Transition to a Market Economy: Lessons from the IMF's Experience," *IMF Survey* (26 July 1993): 239.
35. IMF, *World Economic Outlook October 1994* (Washington, D.C.: IMF, October 1994): 66.
36. Bess Brown, "Election in Kazahkstan" and "Kazahkstan Election Censured," *Radio Free Europe/Radio Liberty Research Report* (7–11 March 1994): 8.
37. Nichol, "Kazakhstan," 3.
38. Ibid.
39. Assan Nougmanou, "Kazakhstan's Challenges: The Case of a Central Asian Nation in Transition," *Harvard International Review* (Spring 1993): 12.
40. Paul Henze, "Turkestan Rising," *The Wilson Quarterly* (Summer 1992): 53. Craig Mellows notes: "The reason for Kazahkstan's apparent order is that only one person really matters: president Nursultan Nazarbayev." He adds: "Nazarbayev combines bold vision with intelligence." In "Kazahkstan: The Big Rush for Gradualism," *Central European* (December 1993/January 1994): 43.
41. Anthony Robinson, "Ex-Soviet Bloc Attracts $42 billion," *Financial Times* (28 September 1993): 2.
42. Frank McGurty, "Philip Morris Expands Into Kazakhstan," *Financial Times* (27 September 1993): 21.
43. Carl Goldstein, "Wells of Hope: Kazakhstan Signs Massive Oil Deal," *Far Eastern Economic Review* (22 April 1993): 76.
44. Development Planning Division, ESCAP Secretariat, "The Asian Republics of the Former Union of Soviet Socialist Republics," *Economic Bulletin for Asia and the Pacific* (June/December 1991): 9.
45. Yalman Onaran, "Transition Proves Hard for Ex-Soviet Republics," *The Christian Science Monitor* (18 November 1992): 9.
46. Arun P. Elhance, "Central Asia's Looming Water Wars," *The Christian Science Monitor* (11 January 1993): 19.
47. European Bank for Reconstruction and Development, *Annual Economic Review 1992,* 117.
48. Nougmanov, "Kazakhstan," *Harvard International Review,* 10.

References

Vietnam

"Foreigners Set Up Shop." *The Banker* (January 1993).

Frank, David. "Vietnam: The Great Leap Sideways." *Euromoney* (June 1992).

Freeman, Nick. "Birth of a Tiger Cub." *The Banker* (June 1993).

Hiebert, Murray. "Do It Yourself." *Far Eastern Economic Review* (31 March 1994): 62.

______. "First Things First." *Far Eastern Economic Review* (11 July 1993).

Kane, Sara. "Transition to a Market Economy: Lessons from the IMF's Experience." *IMF Survey* (26 July 1993).

Kim, Sam Y. "Vietnam's Return to Capitalism: The Next Economic Miracle in Asia?" *Harvard International Review* (Fall 1992).
"Vietnam—Oil Field Awarded." *Far Eastern Economic Review* (14 January 1993).
"Vietnam: Still Divided." *The Economist* (4 April 1992).
"Vietnam's Government Makes Bonds Available to Foreign Investors." *Asian Wall Street Journal* (2 August 1994).
"Vietnam's Struggle for Good Science on the Cheap." *The Economist* (18 April 1992).
Wain, Barry. "Vietnam, Put Off by Privatization Style of Eastern Europe, Takes a Slower Road." *The Asian Wall Street Journal Weekly* (24 August 1992).

Slovenia

Centre for International Cooperation and Development. *Slovenia: New European Partner.* Ljubljana, Slovenia: Centre for International Cooperation and Development, 1992.
European Bank for Reconstruction and Development. *Annual Economic Review 1992.* London: European Bank for Reconstruction and Development, 1993.
The Ministry of Science and Technology of the Republic of Slovenia. *Science in Slovenia: Overview with Highlights.* Ljubljana, Slovenia, 1992.
RFE/RE News Briefs (11-15 April 1994).
Robinson, Anthony and Laura Silber. "Now the Economy Has to Adjust." *Financial Times Survey Slovenia* (30 March 1993).
______. "Republic of Slovenia." *Financial Times Survey.*
Thompson, Mark. *A Paper House: The Ending of Yugoslavia.* New York: Pantheon Books, 1992.
______. "Slovenia—Out of the Maelstrom." *Business Europa* (January/February 1993).
Urban, Rob and Laura Zelenko. "Europe Today: Balkans War Boosts Slovenia, on Battle's Border." *Bloomberg Newswire* (11 October 1993).

Kazakhstan

Barclays ABECOR Country Report. *Central Asian Republics* (June 1993).
Brown, Bess. "Election in Kazahkstan." *Radio Free Europe/Radio Liberty Research Report* (7-11 March 1994): 8.
______. "Kazahkstan Election Censured." *Radio Free Europe/Radio Liberty Research Report* (7-11 March 1994) 8.
Development Planning Division, ESCAP Secretariat. "The Asian Republics of the Former Union of Soviet Socialist Republics." *Economic Bulletin for Asia and the Pacific* (June/December 1991).
Elhance, Arun P. "Central Asia's Looming Water Wars." *The Christian Science Monitor* (11 January 1993).
European Bank for Reconstruction and Development. *Annual Economic Review 1992.*
Goldstein, Carl. "Wells of Hope: Kazakhstan Signs Massive Oil Deal." *Far Eastern Economic Review* (22 April 1993).
Henze, Paul. "Turkestan Rising." *The Wilson Quarterly* (Summer 1992).
McGurty, Frank. "Philip Morris Expands Into Kazakhstan." *Financial Times* (27 September 1993).

Mellows, Craig. "Kazahkstan: The Big Rush for Gradualism." *Central European* (December 1993/January 1994).

Nichol, Jim. "Kazakhstan: Basic Facts." *CRS Report for Congress* (26 June 1992).

Nougmanou, Assan. "Kazakhstan's Challenges: The Case of a Central Asian Nation in Transition." *Harvard International Review* (Spring 1993).

Onaran, Yalman. "Transition Proves Hard for Ex-Soviet Republics." *The Christian Science Monitor* (18 November 1992).

Poujol, Catherine and Pierre Gentelle, eds. *Asie Centrale: Aux Confines des Empires, Réveil et Tumulte.* Paris: Editions Autrement, 1992.

Robinson, Anthony. "Ex-Soviet Bloc Attracts $42 Billion." *Financial Times* (28 September 1993).

"Transition to a Market Economy: Lessons from the IMF's Experience." *IMF Survey* (26 July 1993).

Part III

Old Elephants

7

An Old Elephant: Brazil

According to cynics, Brazil is the country of the future—and it always will be. Although Brazil's 1994 World Cup soccer championship victory over Italy reinforced the country's sense of destiny, many nagging doubts remain about the country's future. However, others have come to hope that Brazil's time in the sun is finally at hand, judging from the rush of investors into Brazilian debt paper traded on the secondary market and even projects since 1991. This process was accelerated by the hopes of a resolution to Brazil's lingering and painful debt crisis with the mid-1992 preliminary commercial bank agreement, which carried on into 1994. The election of a centrist as president in 1994 was also encouraging. Even before these milestones, however, investors were excited about Brazil. Foreign direct investment during the first four months of 1992 totalled $1.4 billion, more than all of 1991, while the Sao Paulo Stock Exchange's Bovespa index soared by 90 percent in dollar terms during the same period. Total foreign capital inflows during 1991 reached nearly $10 billion (although direct investment at $1.3 billion was not impressive). A Eurobond issued by Petrobras marked the country's return to international capital markets, while the first investment trust specializing in Brazilian shares was floated on the London market in mid-1992. During 1993, total foreign investment was an astounding $32.6 billion, while the stock market was booming.

This is progress indeed for a country that until 1990 was generally considered one of Latin America's many hopeless causes despite its strong internationalized private sector. However, the reasons behind this rush into Brazil are decidedly murky. For the most part, investor interest in Brazil has been sparked by a fear of missing a boom market—much like Peru—rather than a deep faith in Brazil's future. This explains why nearly 90 percent of the 1991 capital inflow, for instance, was in the form of short-term, speculative investments rather than bricks and mortar. Lending to Brazilian companies at high interest rates and investing in the undervalued stock market could earn a quick buck, but long-term investment is still scanty and cautious. As the *Financial Times* duly noted, "having seen the turnaround in Mexico, Venezuela, and Chile, investors now fear being left out of, potentially, the next Latin American success story."[1]

Optimists' claims that Brazil is now poised for take-off seem to be based more on hope than reality. Conventional wisdom at one point held that Brazil would mimic the performance of those two Latin American superstars, Mexico and Chile. As one analyst observed: "Brazil was riding on the back of the other Latin American markets. Many investors reasoned, without stopping to make an in-depth analysis, that Brazil would experience the next stock market boom in Latin America."[2] Brazil was essentially hitching a ride, benefiting from the region's improved image largely thanks to moves in other counties where structural reforms were more advanced and inflation more modest. "The enigma of Brazil's market is that it began catching foreign eyes before the country had completed the economic house cleaning that other counties in the region undertook at an earlier stage in their stock market development."[3]

The final verdict is not yet in, and will not be in for many years to come. However, the corruption scandal that eventually toppled President Fernando Collor de Mello in late 1992 was a first and worrisome warning to investors jumping into Brazil. Much like Peru, investors who dove in have been quickly dismayed by stock market volatility and corruption in local politics. The stock market roller coaster of the 1991–93 period displays this trend, as a deepening domestic recession and confusion over privatization took a heavy toll. Such setbacks may be short-term glitches on the road to reform, or symbols of deeper and more lasting troubles. Thus, the fundamental question remains open: Will Brazil always be the country of the future, or is it finally poised for take-off?

Technological Nationalism

As always, we will approach this question through an examination of our variables for success. With regard to technological potential, the country displays some characteristics that are typical of the region, and some peculiarly Brazilian quirks. As in the other developing nations in the area, the rhetoric aside, the government simply does not have enough money to spend on feeding and housing, let alone educating, its armies of young people. (Some 36 percent of the population is under age fifteen as opposed to, say, 21 percent for Canada.) The literacy rate is unimpressive at 76 percent, well below Mexico at 88 percent, and even slightly below Peru (79 percent).[4] This has helped create a huge, virtually unemployable underclass with severe social consequences. The government is under fire for widely publicized murders of street children by death squads. During his abortive tenure in office, the Collor government insisted that the welfare of children and education were high priorities. "We cannot enter the First World with Third World labor. Cheap manpower is no longer an incentive to investment," former President Collor noted. We could not have said it better ourselves—but both the will and the ability to rectify this situation are sorely lacking.

In the meantime, Brazil's labor pool of poorly educated, poorly fed unemployables will not equip the country for world-class economic competition in the 1990s. Much of the burden will fall on a middle class that has been shrinking under the weight of the country's ongoing economic crisis and deteriorating social conditions. To exacerbate the problem, economic and political upheavals have tempted some of Brazil's brightest and best into leaving. Rising emigration suggests that the country may be losing much of its brains and talents, further compromising its chances of emerging from the economic crisis. Most ominously, emigration seems to be hitting the academic and business elite especially hard. So few recipients of overseas scholarships return home that the National Council for Scientific and Technological Development reportedly provides financial incentives to attract them. The effects of this study brain drain on Brazil's future ability to compete are virtually incalculable.

Brain drain and poor education resources are, to some extent, endemic in countries that are struggling to develop. The third factor, economic nationalism, is peculiarly Brazilian in this context but may have severely damaged the country's ability to absorb and use the latest technology. It was embodied in the country's so-called Informatics Law, passed in 1984 to foster a home-grown electronics industry. Instead, the law has engendered an uncompetitive and technologically outdated industry. The law established a "market reserve" in numerous key areas, closing its markets to foreign electronic goods and fostering the growth of a high-cost, low-tech domestic industry. The result: Brazilian-made computer equipment on average costs two-and-a-half times what it would cost in the United States.[5] According to a former science minister, protectionism in this area means that Brazil is now probably a generation behind in computer technology. Even more damaging, Brazil has ended up a "largely computer-unfriendly nation." A recent study found that only 0.5 percent of Brazilian classrooms are equipped with computers, compared to 96 percent in the U.S., while only 12 percent of small and medium-sized Brazilian companies are at least partly automated, compared to 90 percent in the U.S.

The damaging effects of this in the technology-based 1990s are potentially enormous. Belatedly recognizing the need to join the rest of the world, Collor moved to rescind the market reserve and other controversial aspects of the Informatics Law in October, 1992. "By lifting the market reserve, Brazil is acknowledging that closing its markets to foreign electronics goods has also closed off the country to major technological innovations and new products, delaying its modernization, adding to production costs and reducing its global competitiveness."[6] Some observers are skeptical as to whether the war against economic nationalists is truly won, however. They point out that import taxes on computers are still very high. Moreover, the new bill contains so many regulations and stipulations that foreign companies could still be prohibited from production and excluded from benefits, such as tax exemptions, that will be available to national companies. They also might be barred from competing for government and state contracts.

For the most part, then, Brazilians still will have to wait before modern, world-class information technology becomes readily available. Brazil is a small market for these goods, so foreign companies will not be rushing to get in the door even now that the law is rescinded. Most will simply wait until import tariffs are lowered and then export to Brazil. Thus, even in the post-Informatics Law era, Brazil is unlikely to host a strong domestic information technology sector.

All of this adds up to a relatively unattractive score in technology preparedness for Brazil. Things should improve in the future to some extent with the phasing out of the market reserve. Moreover, the commercial bank debt agreement could result in slightly more money being available for education and investment in human capital. However, prolonged delays in concluding such an accord plus the collapse of Collor suggest that lasting reform may be as elusive as ever. In the meantime, Brazil's educational structure (the country is the world's fifth largest by population, but fifty-second in per capita GNP and seventy-fourth in educational national achievement) does not support technology-based takeoff.

Ethnic Diversity

The poverty-stricken lower classes enjoyed virtually no benefits from the "economic miracle" of the 1970s, and they have suffered disproportionately from Brazil's recent economic travails. About one-fourth of the population is illiterate, and Brazil has one of the world's most uneven income distributions, as 60 percent of the national wealth is concentrated in the hands of 1 percent of the population. Chronic inflation has made this problem worse, since the wealthy receive the benefits of extensive indexation, which is not available to the poor. Critics charge that Brazil is really two nations in terms of living standards: Switzerland and Ethiopia. The best-known model in this regard characterizes Brazil as "Belindia," a nation in which 10–15 million people (out of a total population of 150 million) live in a Belgian-style, consumer-oriented, modern, developed economy. Around 100 million reside in an Indialike country, underdeveloped, neglected, unorganized, and leaderless.[7]

The extreme imbalance of income has led to a high incidence of crime, ranging from kidnappings in Rio to land invasions in the rural northeast, posing major challenges for the government. The huge disparity in living standards between the rural northeast and the developed south has produced massive migration to city slums, as well as an incipient southern separatist movement. This has spawned a sharp increase in urban violence, marked by the formation of neighborhood groups that routinely kill assailants, and by retaliatory police brutality. Drug trafficking has found this lawless environment fertile soil. Brazil's position as a transit point for cocaine and heroin has become a problem. Hundreds of street children have been murdered in Rio, allegedly by death squads financed by businessmen and hoteliers trying to clean up the city and reduce petty crime.

While social turmoil is class based, so far it has lacked the ethnic or racial undertones present in Peru. Decades of immigration have made Brazil one of the world's most ethnically diverse nations—nearly half the population is black or mulatto. Slavery was abolished in 1888 and, until quite recently, Brazil was widely viewed as a country with relatively cordial race relations. However, the end of military rule has coincided with a slow but steady transformation in the Brazilian racial dynamic. As one observer noted, "Racial conflict and mobilization, long almost entirely absent from the Brazilian scene, are reappearing."[8]

Celebrations marking the one-hundredth anniversary of abolition were marred by demonstrations and other actions by Afro-Brazilians repudiating what they dubbed "the farce of abolition." These incidents reflect the indisputable fact that deep divisions exist between whites and blacks. A century after the abolition of slavery, blacks lack adequate political representation, housing, and health care. In addition, their living standards are on average far below the standards of white Brazilians. Research over the past four decades has confirmed the centrality of race in Brazilian development, not only past and present, but also in the future. A series of studies sponsored by UNESCO in the 1950s first pinpointed these issues, especially major works by Florestan Fernandes, who definitively distinguished race from class conflict. Since then, overwhelming evidence establishes that "substantial racial inequality may be observed in levels of income, employment, and returns to schooling, in access to education and literacy rates, in health care, in housing and, importantly, by region."[9] Blacks are still overwhelmingly concentrated in the bottom strata, and race is still a crucial determinant of economic success.

Furthermore, the evidence suggests that economic advances in the past two decades have widened, rather than reduced, the gaps between white and black in Brazil. According to one study comparing racial inequality in the U.S. to Brazil, while most measures of inequality in the U.S. declined during 1950–1980, the same measures in Brazil were stable or even increasing. As a result, by 1980 the U.S. was more racially equal than Brazil, partly due to the character of economic growth and specifically the income-concentrating effects of expansion in Brazil. The benefits of growth accrue disproportionately to the upper and middle classes, which are overwhelmingly white.[10]

Despite these problems, Brazil has experienced remarkably little major political turmoil (such as coups, revolution, or civil war) over the past decade in the face of massive economic dislocation. The country has a long tradition of achieving change through compromise rather than conflict. The economic crisis, however, has brought outbreaks of unrest reflecting the heavy costs of austerity programs and rising tensions. These incidents are usually in the form of strikes, demonstrations, or rural disturbances, as well as an upsurge in urban crime. Violent protests greeted the first privatization in late 1991, bus strikes in Sao Paolo have disrupted travel, and supermarkets have been ransacked in Rio.

Considering the severity of the country's economic travails, however, levels of social turmoil have been remarkably low. There is no threat to government stability, and little cost to business from the occasional strikes and demonstrations. Brazil's tradition of peaceful change suggests that social and political tensions will be held in check. However, the presence of deep racial divisions does hold out the potential for severe disruptions in the future.

Many factors in Brazilian history point to a more gradual politicization of racial tensions than in the U.S. There was no major national conflict over racial slavery like the U.S. Civil War, and there has been far less state-enforced racial segregation in Brazil. As a result, most observers expect "no explosive racial upheaval" in Brazil.[11] However, at the same time the political *abertura*, or opening, has permitted blacks to mobilize and organize as they gain in political awareness. Brazil's traditionally harmonious race relations will probably deteriorate as economic growth and modernization create opportunities for upward mobility, which will in turn create fierce competition among Brazilians to seize these advantages. The existence of subtle and flexible forms of discrimination will effectively hinder the access of darker-skinned people to social and economic advancement.[12] As a result, worsening inequality and dashed expectations related to racial tensions could create a volatile mixture indeed.

Wealth in the Ground?

Natural resources are conventionally cited as Brazil's greatest source of potential wealth. The country's enormous buried resources make it one of the world's leading exporters of natural resources and commodities. This has enabled it to run a huge trade surplus, at times second only to Japan in world standings.

However, it is discouraging that despite this huge trade surplus the country has great difficulty keeping its current account out of the red, suggesting that raw materials exports are not sufficient to finance the country's other needs. In fact, the vast minerals sector is severely undercapitalized and intensely dependent on world price movements. Investors are losing interest in the mining sector, as a sharp decline in spending on the quest for mineral deposits indicates. Investment in minerals research and prospecting was estimated at only $40 million in 1992, barely one-third of the 1988 level. (This compares to around $1 billion per year in Canada, a country of comparable size and mining potential.) Given an average lead time of eight years until exploitation, this could presage a severe plunge in mining output over the next decade. This depressed investment climate reflects domestic recession as well as low international prices for most minerals in recent years. Moreover, investors are dismayed by the 1988 constitution, which prohibits foreigners from holding a majority interest in mining operations in many cases. President Cardoso hopes to revise this constitution, but will face obstacles in the Congress. The bottom line: Insufficient investment and vulnerability to low international prices suggest that the mining sector will hardly be an engine of growth for Brazil in the 1990s.

Surprisingly for such a vast country (the world's fifth largest), the agricultural sector is another source of weakness. Brazil is consistently among the world's top three agricultural exporters, but two-thirds of its population have an inadequate daily calorie intake according to the World Health Organization. One-half of the 20 million dwelling in the rural northeast are suffering from some form of malnutrition. The vast agricultural sector is heavily dependent on government support and unstable export markets, and has attracted little interest on the part of foreign investors.

These facts paint an unexpectedly disappointing picture of Brazil's natural resources sector. Not really able to feed itself and still dependent on oil imports (although the amounts are declining as a result of production and substitution programs), Brazil has relied for too long on its fabled minerals wealth. In fact, the minerals sector is undercapitalized and has poor prospects for development given world price trends. Vast mismanagement of the agricultural sector and a declining global role for raw materials in the next decade suggest that Brazil will draw less-than-expected strength from its natural resources in the 1990s.

Sequencing: Second Things First

Issues of sequencing and economic reform are inextricably linked in Brazil. The country's emergence from military rule was hailed worldwide in the 1980s, as it progressed slowly but surely toward democracy.

Collor took office in 1990 as the first directly elected president in thirty years, a move widely viewed as marking the end of Brazil's passage to democracy. However, the democratic system that succeeds the generals is unsophisticated and fragile—nothing illustrates this point more dramatically than the summary removal of Collor from office in late 1992 for alleged corruption. Two decades of authoritarianism have resulted in an absence of strong and responsible democratic institutions. Few Brazilians associate themselves with any of the thirty-two registered political parties, which are perceived as opportunistic and lacking in clear goals and ideology. Rather than building a new set of institutions, democracy has unleashed a new sort of forces allied only against reform and progress.

The 1988 constitution is an important obstacle to reform. The document provides for a fundamental transfer of power from the executive to Congress, specifically abolishing the decree laws that allowed the President to rule almost unchecked. Congress must approve all legislation, including major budgetary and economic decisions. The constitution is contradictory and confusing in places, enshrining important civil and labor rights while sanctioning the power of the military and landowners. The constitution invites the government and Congress to battle for power, and the judiciary is asked to settle many disputes. Moreover, it provides the state governments greater control over their finances—a practice that has been widely abused for political reasons. President Cardoso has launched efforts to revise the constitution, and the outcome of this process will be crucial in defining the investment climate for years to come. However, the power of the forces ranged against economic and political reform decidedly dim the prospects for substantial constitutional reform.

Operating under a constitution that made him Brazil's weakest head of state ever, Collor was forced to rely heavily on congressional support. The Congress elected in October, 1990 was predominantly center-right, but Collor's own National Reconstruction Party controlled only about 8 percent of the seats. Collor's downfall reflected his inability to control the levers of power, and his successor, the lame-duck President Itamar Franco, witnessed a continued erosion of public support and executive power. The Congress has on balance proved a deeply conservative force in the struggle to reform the Brazilian economy, focused on maintaining entrenched privileges and subsidies rather than on creating a free-market environment. The new Congress elected in 1994 appears to be more reformist, but

Cardoso will encounter many obstacles as his reforms begin to threaten vested interests. While other Latin American leaders were able to rely on a one-party state (as in Mexico), a strong Congressional majority, or a dictatorship (Chile) to push through reform, Brazil's Congress is so fragmented that it contains fully nineteen parties. Sadly, corruption is another factor—a mortal one in undermining the Collor administrations's drive to modernize the economy. An even greater corruption scandal emerged in early 1994, highlighting rampant disillusionment with the political system and even with democracy itself.

Labor has emerged as another fundamentally conservative element in the fight on reform. Collor hoped to shake up the cosseted, inefficient, and state-dependent private sector by reducing tariffs and other measures, which to some extent he did before his ouster in 1992. However, progress was thwarted by his inability to break the political stalemate created largely by labor leaders. The issue of port modernization is a perfect example. Exorbitant handling costs due to union control rank Brazilian ports among the most expensive in the world for imports and exports. Collor's port modernization bill was bottled up in Congress for well over a year due to intensive lobbying by organized labor. The bill that finally emerged was radically different from the original government proposals, which sought to privatize and modernize the ports while breaking the unions' right grip. Instead, the compromise bill allows for only partial privatization and some limited modernization, but retains key union privileges. This is a luxury that Brazil will not be able to afford; as one manufacturer warns, "We are afraid that if the port obstacles continue, our exporters will not be able to compete with the Asian 'tigers.'"[13]

The ports are a good example of how powerful forces ranged against Collor largely frustrated his drive for reform. Labor is one strong group. Another is the employers organization FIESP (which is usually referred to as "the powerful FIESP," as if this were its full title). The bureaucracy and labor unions oppose privatization and job losses in the public sector, while business and farming interests are dismayed by the loss of fiscal privileges and subsidies, as well as by import liberalization. In addition, the armed forces remain a key political power confirmed by the constitution as the ultimate guardians of law and order. Military leaders are deeply concerned over low levels of pay in their ranks, charging recently that a fighter pilot earns the same as an elevator operator in Congress. Even a free press worked to undermine Collor. "Ironically, the extensive press freedom under the new democracy has worked against Mr. Collor,"[14] and inadvertently helped to bring him down.

Of course, this is not ironic but merely predictable. Like Fujimori in Peru and the rebelling officers in Venezuela, Collor ultimately became disillusioned with Brazil's political institutions, realizing that economic reform was much easier before voters and interest groups appeared. As Brazil's first directly elected president in three decades, Collor expected to go down in history as the man who saved the Brazilian economy. Mired in a messy corruption scandal and seeing his reform plans checked at every turn, Collor was eventually forced to resign at least in part by the forces unleashed by democracy. He became at least as much a "victim of his own democratic reforms"[15] as of his own greed and mismanagement. Ironically, it was corruption of the old order—which he had vowed to clean up—that caused Collor to fail. Unfortunately, he appeared to many Brazilians to be part of the same old political system that existed prior to the young leader's entrance on the presidential stage. His succession by Vice President Itamar Franco in October, 1992 was hardly inspiring for Brazil's future, considering the new chief executive's track record as an economic nationalist with little comprehension of economics. As one Brazilian business leader lamented to the authors in December, 1992: "Collor was a dishonest man with the right ideas; Itamar is an honest man with the wrong ideas." To this another businessman added: "Collor moved in the right direction. We are on the runway, ready to fly, but with Itamar we may end up stuck on that runway." By mid-1994, these observations proved largely accurate—the Franco administration had gone through four Finance Ministers, inflation remained over 40 percent a month (before falling to single digits in the late Summer), and the deal with the commercial banks over the country's external debt remained unconsummated. President Cardoso is a much more encouraging figure, but the strength of the anti-reform forces is still dismaying.

In a historical context, the lack of support for economic reform is not surprising. Brazil's political scene has long been dominated by "an unyielding and greedy elite, staunchly opposed to real power-sharing and determined not to pay taxes." This elite, represented by the power structure outlined above—labor, business, Congress, the military—is a fundamentally conservative bloc ranged against market reforms. "The Brazilian political system is embedded comfortably in a pyramidal social structure that tolerates no real challenge to existing elites and their political system." Under these circumstances, any president is likely to be condemned to "frustration and political impotence." In the end, Brazil's elitist system is its biggest obstacle to economic development.[16]

Reluctant Reformers

As the above discussion suggests, Brazil's will and ability to implement meaningful economic reforms are dubious. Its macroeconomic changes are moving slowly by Latin American standards, leaving the country lagging well behind its neighbors in the reform process. Brazil now has the highest inflation in the region. Indeed, it is the only large South American country still grappling with hyperinflation. It entered 1994 with an inflation rate of 40 percent a month and virtually nonexistent economic growth, compared to average growth in Venezuela, Mexico, Chile, and Argentina of around 5 percent. While Cardoso's Real Plan reduced inflation by year-end 1994, the fundamental causes of inflation remain unchecked. More alarming, it remains the region's most stubbornly closed market. As the rest of Latin America opens to freer trade and extensive deregulation, reducing inflation and creating real growth for the first time in a decade, Brazil is increasingly alone as the country that has yet to put its economic house in order. Import tariffs are the highest among large Latin American countries, at an average 21 percent and a maximum 65 percent, while imports still make up barely 5 percent of GDP. Brazil got a late start on privatization; it was the last major country to reach an accord with the IMF and a definitive resolution of its foreign commercial bank debt problem, and it boasts a minimum salary lower than Bolivia's.

For a time, optimists hoped that, in retrospect, 1991 would eventually come to be seen as the turning point for Brazil when Collor finally succeeded in implanting a liberal agenda. The ban on computer imports was lifted in late 1992, the stock market was opened to foreign investors, and privatization finally got underway. Collor, at least, seemed to accept that the gimmicky economic shock plans that characterized recent policymaking must be discarded in favor of attacking the fiscal deficit as the root cause of inflation. More fundamentally, he began to persuade some people that Brazil could no longer resist the worldwide rush to open markets. The country at long last reached a preliminary Brady Plan debt deal with creditor banks in mid-1992 to reduce foreign debt, provide lower interest rates, and extend the term of its loans. This accord seemed crucial to government plans to liberalize the economy, attract foreign investment, and stabilize prices.

But despite this progress, there were ominous signs even during 1991–92 that economic reform was on shaky ground in Brazil. The country failed for two consecutive quarters in 1992 to meet its macroeconomics targets with the IMF, reflecting in part the intensity of domestic political wrangling over long-term fiscal reform. Brazil's troubled privatization program serves as an instructive symbol of the confusion and disarray still besetting the country. Even after violent protests failed to derail the first sales (though they did manage to force the government into several deeply embarrassing delays), the future of newly privatized concerns remains murky. Leonel Brizola, state governor of Rio and a leading opposition politician, vowed that if elected president in 1994, he will renationalize the former state steel company, Usiminas. Said Brizola, "[T]he foreign group that gets their hands on Usiminas should know that they will have to face some Brazilians down the road."[17] Not surprisingly, foreigners bid for only 6 percent of the company's shares.

While Brizola and other leftists were soundly defeated in the 1994 elections, the country's fundamental commitment to economic reform is still suspect. The contretemps over privatization was an early warning sign of Collor's apparent inability, amid political opposition, to carry out meaningful reforms. The foreign debt deal does solve some problems, but even more serious problems linger at home. After decades of economic nationalism, the state-run model is accepted throughout the economy. This has created huge and powerful constituencies threatened by its removal—the elites referred to above. Business is cartelized and uncooperative, often lobbying successfully to slow or even halt the liberalization process. The powerful São Paulo Federation of Industries (FIESP) prospered from import substitution policies of the 1950s through the 1970s. Today it houses many "dinosaurs" unwilling to allow outsiders into one of the world's most protected economies, and is frequently at odds with the government as a result.

Everything comes down to two critical variables in the end: fiscal reform and responsible government leadership. The economy cannot be stabilized and inflation controlled without a thorough housecleaning of the government books. The government's domestic liabilities created debt service costs of fully $21 billion in 1992, or nearly 5 percent of

GDP. Unless interest rates decline, the government will be unable to service its own domestic debt in the very near future. Cardoso's Real Plan reduced inflation dramatically during 1994, but largely at the cost of an increasingly overvalued currency—which is coming under attack in the wake of Mexico's troubles. The president still has to attack the real root causes of inflation; the hard work lies ahead.

Congressional action is urgently needed for a permanent solution to reform and simplify the tax laws, delineating more clearly the division of responsibilities between federal and state governments. At present, the federal government must pass on almost one-half of its tax revenue to the states and municipalities, who are notorious free spenders. Companies must cope with more than fifty separate taxes; the extreme complexity of this system helps to fuel a flourishing parallel economy off the tax books, and massive additional losses of revenue to the state. The constitution also prohibits the government from sacking federal employees, an obvious obstacle in the drive to streamline government operations. Given these problems, it is hardly surprising that Brazil consistently fails to meet IMF targets on government operations.

The solution to these problems brings us full circle to sequencing issues once more. Constitutional changes are required to overhaul the tax system, permit sacking of state employees, and shift responsibilities between the state and federal governments. However, this in turn requires a three-fifths majority in a Congress that just threw out one president and subsequently coexisted with a lame-duck, unelected one. Cardoso may be able to command a three-fifths majority through de facto coalitions, but this support is tenuous. Without constitutional changes, fiscal consolidation will not be possible. Without fiscal consolidation, inflation cannot be permanently tamed. Without price stability, growth will not resume.

Conclusion

This bleak outlook suggests that while Collor and Cardoso may have succeeded in implanting the idea of economic reform, the strength of the forces ranged against modernization makes success questionable. Much like Peru, the weakness of Brazil's democratic institutions is stifling the president's ability to implement economic reform. In Brazil, the dinosaurs opposing the president are especially strong. Endemic corruption, which even came to embroil Collor and his family, saps the strength and erodes the credibility of the government. Unlike in Mexico and even Argentina, substantial sections of the Brazilian business community are deeply threatened by liberalization and continue to be a conservative force in the years ahead. In this struggle business will have a strange bedfellow in labor, also threatened by liberalization and emboldened by its close ties to national politics. Finally, the existence of 1988's Alice-in-Wonderland constitution—with its wide-ranging contradictions—will impede serious economic progress. The danger for Brazil is that the failure of economic reform will eventually threaten the survival of democracy and force the country's leadership to adopt some form of soft or hard authoritarianism—which, sadly, is not the ultimate answer.

Notes

1. *The Financial Times* (2 October 1991).
2. Ibid. (27 November 1991).
3. Ibid. (29 May 1992).
4. Political Risk Services, *Country Forecasts: Brazil* (Syracuse, New York: Political Risk Services, June 1993).
5. *The Wall Street Journal* (8 August 1991).
6. Ibid.
7. Brady Tyson, "Brazil Today: A Study in Frustrated Democratization," in Jan Knippers Black, ed., *Latin America, Its Problems and its Promise* (Boulder, Colo.: Westview Press, 1991, 2nd ed.), 564.
8. Howard Winat, "Rethinking Race in Brazil," *Journal of Latin American Studies 24* (February 1992): 173–78.
9. Ibid.
10. George Reid Andrews, "Racial Inequality in Brazil and the United States: A Statistical Comparison," *Journal of Social History* (Pittsburgh, Pa.: Carnegie-Mellon University Press, Winter 1992): 254.
11. Winant, "Rethinking Race in Brazil," 191.
12. Andrews, "Racial Inequality in Brazil and the United States," 229–54.
13. *The New York Times* (25 June 1991).
14. *The Financial Times* (3 July 1992).
15. Ibid.
16. Tyson, "Brazil Today," 574.
17. *The New York Times* (24 October 1991).

References

Andrews, George Reid. "Racial Inequality in Brazil and the United States: A Statistical Comparison." *Journal of Social History*. Pittsburgh, Pa.: Carnegie-Mellon University Press (Winter 1992).

Black, Jan Knippers, ed. *Latin America, Its Problems and its Promise, 2nd edition.* Boulder, Colo.: Westview Press, 1991.

The Financial Times (2 October 1991).

______. (27 November 1991).

______. (29 May 1992).

______. (3 July 1992).

Maybury-Lewis, Daniel. "Brazil's Significant Minority." *The Wilson Quarterly* (Summer 1990).

The New York Times (25 June 1991).

______. (24 October 1991).

Political Risk Services. *Country Forecasts: Brazil.* Syracuse, New York: Political Risk Services, June 1993.

The Wall Street Journal (8 August 1991).

Winant, Howard. "Rethinking Race in Brazil." *Journal of Latin American Studies 24* (February 1992): 173–92.

8

Peru and Venezuela

Peru and Venezuela in Latin America represent two cases of countries caught on the elephant track. Both nations, like Brazil, have considerable potential, but a number of factors are missing that hinder their development.

Before the *autogolpe,* or self-generated coup initiated by President Alberto Fujimori of Peru in April, 1992, this South American country of 22 million people was experiencing an upsurge in foreign investor interest. "Investors seeking the next Latin American success story have been flocking to Lima," noted the financial press.[1] A flurry of overseas funds had driven the stock market up by 30 percent per month during the last quarter of 1991, as analysts chased the "great opportunity" of Peruvian economic reform. According to one broker: "Investors love a back-from-disaster story. They are looking for a repeat of the kind of gain, about 300 percent a year, made in Argentina."[2]

Such hopes were encouraged, even fanned, by the ever-optimistic Fujimori. During a sales trip to Asia in late 1991, he touted Peru as "the ideal scenario for investors who do not flinch before some difficulties." Peru's bright future as a Latin American "jaguar"—following in the footsteps of Asian tigers—was stressed over and over by Fujimori. Businessmen were beginning to respond. A flood of speculative capital (augmenting, no doubt, the ever-present inflows of drug money) had made Lima one of the most expensive cities in Latin America. More positively, major copper mining investments were underway and oil exploration deals were also in the pipeline.

All of this progress came to a sudden, jarring, and perhaps brief halt on 5 April 1992, when Fujimori suspended the constitution, dissolved congress, imposed press censorship, and arrested leading political opponents. In the first few weeks after the autogolpe, at least $200 million in capital fled the country and bank deposits were decimated. Even more damaging, United States-led international opposition to Fujimori's attack on democracy resulted in a temporary cutoff in foreign aid and credits. It appeared that Peru was heading back to the abyss of international isolation from which it had just emerged under Fujimori's leadership. While the resumption of a more democratic track and steady progress on economic reform have allayed these fears, doubts on long-term economic prospects remain.

Peru's latest difficulties are only magnified by a glance at its potential. Thirty years ago (1963–93) Peru and South Korea had the same per capita income levels. Today Peru's per capita income is barely at 1960s levels, while South Korea "sends aid."[3] Peru's economy today is characterized by minimal development levels despite abundant mineral resources. The only sure flow of funds is from the lucrative and booming illicit drug trade, which brings an estimated $1 billion per month into the banking system. The downside: Drug money keeps the monetary system liquid but overvalues the currency, making the cost of living high and exports uncompetitive.

For true believers, this is just another opportunity—more golden than ever—to buy into Peru on the cheap. With the stock market depressed following Fujimori's coup and few but the hardiest investors looking seriously at Peru, that may have been an ideal time to invest in its long-term potential. Indeed, barely one year after the coup *Euromoney* reported that "bolder investors and entrepreneurs with an eye for a bargain are flocking to Peru." The big names rushing into Peru include a blue-chip list of investment and commercial banks, major consumer products companies, and oil enterprises. Businessmen from Chile, Colombia, and China are looking into the potentially rich but underdeveloped, cash-starved fishing and mining sectors, contributing to the "biggest burst of investor enthusiasm for years."[4] In 1992–93, foreign investment surged to $2 billion in sharp contrast to the 1985–90 period when virtually no new investment arrived. Big-name companies are attracted by the privatization program and by the maturing stock market. The business climate has been enhanced by new laws to promote foreign investment, making Peru arguably the most liberal environment for foreign capital in Latin America. This positive feeling was reinforced by the arrest of two major guerrilla leaders in late 1992, the more significant of whom was Abimael Guzman, head of the Maoist Sendero Luminoso, or Shining Path. But does Peru have the potential to become a Latin American jaguar? Fujimori's faith notwithstanding, our model raises serious doubts about long-term profitability in the Peruvian economy from the viewpoint of foreign investors.

But Seriously, Folks

Leaving aside the autogolpe for the moment, Peru under Fujimori in some ways appears to be making all the right moves. When Fujimori first took office in mid-1990, he inherited a country that was not on speaking terms with the international financial community.[5] The economy had contracted by an estimated 30 percent between 1987 and 1991, and had been battered by terrorism, capital flight, and hyperinflation. When Fujimori took over, inflation was soaring to 6,650 percent for the year, some 80 percent of the work force was unemployed or underemployed, and one-third of the population was malnourished. Symptomatic of these woes, a cholera epidemic in 1991 eventually cost the state over $100 million in medical bills, lost food exports, and foregone tourism income. Moreover, Fujimori had to grapple with a foreign debt of $20 billion and arrears on debt service to multilateral financial institutions of over $2 billion, which made the country ineligible for new aid.

While many of these problems reflected economic mismanagement under Fujimori's predecessor, Alan García, they also illustrate Peru's structural economic ills.[6] García, a populist at heart, stimulated growth through demand-led policies, which discouraged both investment and exports while feeding inflation. García sought to preside over a nationalist, consumer-led "heterodox" economic recovery program. This encompassed a unilateral decision that Peru would limit interest payments to no more than 10 percent of the nation's export earnings, the freezing of prices and exchange rates, the raising of wages, the lowering of taxes, the creation of public employment programs, and the reduction of domestic interest rates.

The results were disastrous. Foreign exchange reserves declined precipitously, and investor and creditor confidence further tumbled, especially following García's ill-timed and ill-conceived move to nationalize banks. The situation was aggravated by Peru's poor standing in the international financial community. García's initial decision in 1985 to limit debt service payments to 10 percent of export receipts was followed by a complete suspension of debt payments. Peru became ineligible for IMF and World Bank credits, placing it at odds with the private international financial community. Private foreign capital inflows dwindled into insignificance.

Fujimori's appearance on the scene, however, breathed new life into the patient. Even before his July, 1990 inauguration, Fujimori reached a basic agreement with the IMF and other development agencies to stabilize the economy, resume payments on some external debts, and reinsert Peru into the international financial system. Upon taking office, he quickly moved to remove long-standing government subsidies and to institute a free market plan drawn up by a team of economists. The plan included measures to cut tariffs, eliminate restrictions on foreign investment, free exchange rates, break up monopolies, and set up conditions for privatization of state enterprises. Fiscal austerity and monetary stringency have sharply restricted money supply growth and reined in inflation.

Fujimori's tough and initially promising program was jeopardized by international reaction to the autogolpe. With hundreds of millions of dollars in foreign aid temporarily frozen, Peru faced yet another balance of payments crisis. Although most aid was restored following elections, creditor governments remain dubious and tentative. Any real economic progress will depend on a full and permanent restoration of international credit lines and hefty inflows of private investment capital. While international pressure on Fujimori has eased, his ambiguous commitment to democracy still raises difficult questions for Western nations, particularly the U.S. President Bill Clinton has opted for flexibility so far, praising Fujimori's crackdown on terrorism and inflation. Other countries are inching toward a grudging acceptance of Fujimori based on his continued economic reforms and certain democratic steps, but they remain wary.

The Technology Gap

Whether this overseas investment capital will be forthcoming, in turn, rests on investors' perceptions of long-term potential in Peru. As noted above, we believe that a crucial success factor for emerging economies in the 1990s and beyond will be their ability to compete in a highly technological world environment. In this context, Peru will be poorly positioned. Its prospects are weakened by extreme poverty, deficient educational systems, and a very low level of technological expertise among the population. Many Peruvians lack basic sanitation, electricity, and water, especially in the Lima shantytowns where such diseases as typhoid and tuberculosis are common. The level of telephone provision is the lowest of any Latin American country, just 2.4 per 100 inhabitants. U.S. emergency food programs feed an estimated 3 million people, or one in seven Peruvians, who would otherwise starve. Infant mortality is high (sixty-seven deaths per thousand), literacy levels are under 80 percent, and nearly 40 percent of the work force is still engaged in agriculture.[7] While tax collection is improving, widespread tax evasion still prevents the government from being able to provide basic health and educational services, especially since the war on terrorism still requires heavy state expenditures. Although Fujimori has recognized the educational deficit and will hike spending in this area where he can, the extreme constraints of Peru's economic situation effectively prevent any real improvement in social indicators over the next decade.

Internal (Dis)Unity

The single greatest obstacle to progress in Peru, however, is its high level of internal strife. As observed above, poverty and class struggle in themselves do not constitute serious obstacles to development (witness the Chilean model for proof). However, when the class struggle is allied with racial and ethnic divisions, it can become a binding constraint (as in Sri Lanka and much of Africa). This, unfortunately, is the case in Peru.

The role of the Sendero Luminoso guerrillas in inhibiting economic development in Peru has been well documented. The government essentially has been fighting an undeclared civil war against these forces, which has claimed more than 25,000 lives in the past decade and cost roughly $23 billion in terms of economic damage (coincidentally, also the size of Peru's foreign debt). Sendero Luminoso guerrillas are strict Maoists. A ruthless and brutal group, they do not try to justify their actions except in long-term objectives of eliminating the Peruvian governmental system and establishing a communal society without private property. The group is actively supported by perhaps 5–10 percent of the population, primarily lower middle-class university students and poor youths from the slums of Lima and the central and southern highlands, but it has controlled up to 40 percent of the countryside.[8] Its leaders tend to be educated middle-class Peruvians from Lima, while its soldiers are usually peasants of Indian extraction.

The Sendero Luminoso rebels do not make direct racial appeals; indeed, leaders insist that they are motivated purely by class struggle, not race. However, independent observers agree that the deep racial divisions of modern Peruvian society play a key role in fueling the anger and xenophobia of the rebellion. "It is impossible to understand Peru's political violence if you only look at poverty or political oppression," notes a leading researcher of the Shining Path movement. "Racial discrimination is a very important factor."[9] Another expert concurs: "It is difficult to understand the acceptance and even support which Sendero Luminoso came to enjoy in the 1980s without first understanding the strong sense of ethnic and cultural marginalistion in the sierra. Indeed, in Peru, as elsewhere, a strong sense of ethnic identity has been one of the factors that has favored rural insurgency."[10]

Peru's long and sad history of racial discrimination is beyond doubt. After Pizarro founded Lima, the Spanish adopted racial controls to maintain their hold on power, barring Indians from owning horses, wearing Spanish dress, bearing arms, and even entering cathedrals. The enormous gulf that the brutal sixteenth-century Spanish conquest opened between the victors and the vanquished has never been entirely bridged. The Spanish created a society in which a tiny ruling class and their descendants came to utterly dominate the Indian, mestizo, and black majority—a historical fact that still haunts Peru today.[11] As late as the mid-nineteenth century, British companies enslaved Indians to work the mines in which thousands died. Today, of Peru's 22 million people, 60 percent are mestizos (people of mixed Spanish and Indian heritage), 30 percent are Andean Indians, and 10 percent are whites. Miss Peru always sports pure Spanish features, while mixed race children play with European-looking dolls.

The Shining Path would probably not exist were it not for Peruvian history of the past four centuries. "Like other political groups throughout Peruvian history, it has cynically exploited what might be called the Indian question," observes one writer.[12] In this environment, it is hardly surprising that arrested Sendero Luminoso supporters often hail from cities of the Andean heartland such as Cuzco and Ayacucho. In fact, Sendero's leaders today are not very different from those who led Indian uprisings against their conquerors in the past. Thus, the guerrilla group capitalizes on the Indian question while aiming its larger message at the increasingly impoverished mestizo underclass of Lima and other cities. One sociologist notes, "Many poor people see whites as foreigners and they have revenge fantasies. In some factories where Sendero has killed white managers, the workers approved."[13]

But if the driving force behind Sendero Luminoso is the racism that permeates Peruvian society, it especially thrives on hunger and despair—and this bodes ill for Peru's future. Partly as a result of Fujimori's tough economic program, fully one-half of the people now live in "extreme poverty" according to the United Nations. Fujimori hopes to bring peace by strengthening a military force known for massive human rights violations, and by arming civilians in the villages. The jailing of Guzman and others has created a leadership crisis within Sendero Luminoso, provoking a sharp decline in morale and financial resources. Desertions have multiplied, and the level of deaths due to political violence during 1993 was halved. But despite these signs of progress, the rebellion has not collapsed. The force of Peru's class conflict—fueled by the rage left by centuries of racial discrimination—suggests that Fujimori's efforts will meet with only mixed results. In turn, poor prospects for social unity in Peru may well doom its chances for economic prosperity.

Sequencing: The Wrong Way Around

Fujimori's autogolpe was another effort to deal with the deep social and political divisions tearing Peru apart. He says—probably quite truthfully—that his actions were taken in order to defend and deepen his economic program, which risked being derailed by conflicts within a corrupt, divisive congress and judiciary. More fundamentally, Fujimori's suspension of democracy reflects his recognition of the reality that it is much easier to implement meaningful economic reform in a closed political system than in a democracy. He has always admired the economic transformation wrought in Chile under General Pinochet.

Indeed, the coup was initially welcomed by most Peruvians, including the business sector, for exactly these reasons. Opinion polls overwhelmingly supported the autogolpe, citing Peru's corrupt judiciary and political parties as the real culprits rather than Fujimori, who retains approval ratings in the 70–80 percent range. Obviously, this was not "just another textbook Latin American coup," as the press duly noted.[14]

Ironically, Fujimori probably agrees with the statement by former president Alan García, who warned (in stating his opposition to IMF-guided economic policies): "You cannot squeeze these poor countries like this and expect to maintain democracy."[15] Fujimori's likely response: Okay then, let's suspend democracy.

In announcing his autogolpe, Fujimori was really only trying to emulate the Asian tiger model: bring the economy up to twenty-first century standards under an authoritarian political system, then turn your attention to political reform. It worked for Korea, Taiwan, Singapore, and even Japan. It worked for Chile and appears to be doing the same for Mexico. Unfortunately for Fujimori, the U.S. and other nations may not let it work for Peru. Creditor nations and multilateral financial institutions, prodded by the U.S., froze more than $1.5 billion in debt relief and development grants in response to Fujimori's announcement. As a result, Peru faced international isolation once more and Fujimori moved to restore democracy, "almost entirely (as) the result of external economic pressures."[16] Fujimori remains a largely authoritarian ruler, backed by the enormous manipulative power of the armed forces. He is a man out of time, attempting to emulate a model that is no longer acceptable in the 1990s new world order. Thanks to U.S. insistence on democracy before economic reform, Peru is likely to get neither.

On the flip side, it is true that Fujimori has managed to get much of the aid flowing back into Peru. He held elections for the November, 1992 constituent assembly and has reestablished some form of democracy or, some would argue, soft authoritarianism. The new constitution gives him the chance to run for a second term, which he will probably win; this would certainly raise questions over whether he has run out of time.

Economic Reform: All the Right Moves

Moreover, it is only fair to say that Fujimori has made all the right moves on economic reform. His tough and initially successful program to maintain discipline over government spending, fight inflation, and privatize dozens of state companies has won worldwide recognition. He shows a remarkable willingness to open Peru to world markets, deregulate the ports, liberalize labor laws, free foreign exchange regulations for businesses to remit profits, and establish arguably the most liberal foreign investment climate in Latin America. No less an authority than

Michel Camdessus, managing director of the IMF, praised the structural adjustment program as "an example for many countries around the world." Similarly, the GATT praises Peru's commitment to abandoning its long-standing import substitution and state intervention model.

However, long-term prospects for Fujimori's program were always shaky, and the world's insistence on return to some form of "democracy" probably sounds a definitive death knell. Fujimori always faced a dilemma: either economic reform fails under a democratic system because of a corrupt and self-interested set of opposition groups, or economic reform fails under an authoritarian government because foreign aid is suspended. Now he is under mounting pressure to raise military expenditures and adopt more populist economic measures to retain the loyalty of the armed forces, as well as the majority of poverty-stricken Peruvians who supported his coup. The return of "democracy," Peruvian-style, will bring a blizzard of self-interested demands and obstacles to IMF-style reform.

The trouble is that neither authoritarianism nor "democracy" are likely to do the job for Peru. A return to authoritarianism is certainly not the answer; by the early 1980s most of Latin America's military governments were wallowing in a mess of failed economic policies and foreign debt crises. Also, elements on that side are at least as corrupt as among the "democrats." There are a number of allegations about links between drug dealers and some of Fujimori's closest security advisors. However, a blind Western insistence on "democracy" that fails to recognize the subtleties and intricacies of Latin American politics is misguided at best.

In addition, Fujimori's reform program, while laudable, adhered too strictly to IMF principles in defiance of Peruvian realities. As a result, here too the sequencing was out of order. First and foremost, Fujimori moved quickly and decisively to liberalize trade and foreign exchange markets—winning the praise of IMF bureaucrats as a result. Someone forgot to test these measures against the Peruvian market, however. An inflexible domestic money market and hyperinflation, coupled with the inevitable influx of drug money, drove interest rates and the currency up to new highs against the dollar. At the same time, domestic manufacturers were crippled by relatively high labor and capital costs, security costs that add at least 10 percent to total charges, and the need to buy certain high-priced local components. As a result, liberalization destroyed a number of viable local companies as well as the deadwood. Nissan, for instance, was forced to close its twenty-one-year-old Peruvian assembly plant while Toyota scaled back to negligible levels, citing their inability to compete without tariff protection given high Peruvian costs and the overvalued currency. These problems illustrate the pitfalls of an inadequate reform package, when the external sector is liberalized first while domestic producers are still crippled by continuing structural rigidities. Even while praising Peru's reforms, the GATT warns that the private sector is slow to respond, blaming high interest rates and taxes, competition from the informal sector, inadequate infrastructure, an overvalued exchange rate, and terrorism.

Natural Resources: All the Wrong Ones

Finally, we come to Peru's much-vaunted natural resources. Much of the recent investment is predicated on Peru's undoubted advantages in this area, both actual and potential. Peru is one of the world's top seven mining countries, with an estimated 7 percent of total international zinc reserves, 10–20 percent of copper and silver, and a long list of other important metals. It ranks second only to Mexico in total silver output, takes fourth place in world zinc rankings, and eighth in copper. Mineral exports bring in more than $1.5 billion per year, almost one-half of total export revenues.

While the mining sector is currently sunk in gloom and depression, investors reasoned that with this much wealth in the ground, Peru's longer-term prospects were rosy. However, our model suggests that traditional mineral resources will be of declining importance in a technology-dominated global macroeconomy. Indeed, the reasons for the "short-term" decline of Peru's mining sector are illuminating. Observers cite several factors for the industry's collapse, most of which are beyond its control: low world mineral prices; the overvalued Peruvian currency; high local costs, including expensive energy and substantial outlays for security; high capital costs and indebtedness; and mismanagement. But what if these are structural facts rather than short-term ones? What if low world mineral costs, economic mismanagement, and political terrorism are not passing fancies, but deeply imbedded trends?

In fact, we would argue that while Peru is rich in mineral resources, these will not be particularly helpful in driving it to success in the 1990s. Low mineral prices will probably be a fact of life, as technology surpasses natural resources as a source of strength. Peru is poor in the resources that will count—energy, food, and a well-educated population. It imports oil (though exploration is underway), and land distribution programs have created serious food production bottlenecks. With a literacy rate of 79 percent according to the World Bank, Peru is also poorly prepared to adapt the technology needed to become a fast-track economy. Peru suffers from a brain drain of skilled people, seeking a better life beyond Peru's troubled borders. Today Peru survives on food imports, with U.S. donations feeding one in seven Peruvians. Peru's most important resource for the 1990s, sadly, may be its ability to produce more than 60 percent of the world's coca leaf, the raw material for cocaine.

Conclusion: A Lizard, Maybe

One writer on Peruvian history notes that the country's problems form a "peculiarly intractable knot."[17] As a result, prospects for the future are lackluster at best. Two glaring weaknesses are likely to prove to be fatal flaws in the future: poorly managed democracy and poverty. First of all, Fujimori's alliance with shadowy and repressive elements in the armed forces shows how much control the army has over the government, and does not bode well for future international relations. The state is particularly susceptible to "clientelism, corruption, and inefficiency," partly due to the lack of any real democratic control over the government.[18] These problems are magnified by the autonomy and unaccountability of the most powerful state institutions, as well as the absence of effective controls against malpractice. Not surprisingly, these factors contribute to the persistence of authoritarian personalist politics, which are now embodied in Fujimori himself. It also undermines the legitimacy of the state in the eyes of the people. His authoritarian tendencies may be growing, as depicted by growing reports of torture, summary trials and executions, even a secret army antisubversive death squad. His image as a benign despot may erode, forcing Peru into international pariah status once more. But if he does loosen his grip on power, the weakness of Peru's "democratic" institutions virtually guarantees substantial backpedaling on economic reform—a perfect Catch-22.

The second fatal flaw is the pervasive and obstinate problem of poverty and ethnic relations. Fujimori's rigid economic policies will, in the short term, create further job losses and reductions in purchasing power.

Shanty towns around Lima are already teeming with despair and resentment, creating rich feeding grounds for the Shining Path. The capture of leader Abimael Guzman in late 1992 was certainly a major blow for the organization, but possibly not a death blow. Development patterns in Peru have done nothing to alter its social and geographical disparities, while its ethnic divide remains one of the worst in Latin America. Fed by these miseries, Peru's guerrilla insurgency is far from defeated. Even if it is fatally wounded, the Shining Path's death throes will be protracted and bloody. As long as deep poverty and racial inequities persist, the Shining Path will attract supporters among the marginalized elements of society.

Venezuela: What a Difference Oil Makes

Venezuela and Peru, although very close both geographically and in terms of raw size, are worlds apart in other ways. Both are located in South America, with roughly equal population levels (Venezuela has around 20 million inhabitants, Peru 22 million), both richly endowed with natural resources, and yet the two countries have developed along very separate paths in the postwar years.

A comparison of 1991 data makes this painfully clear. The differences can be summed up in per capita income data, revealing Venezuela to be far ahead of its southern neighbor with income per head of $2660 to Peru's $1770. Growth in Venezuela during 1991 was a whopping 9.2 percent; Peru limped along at 2.4 percent. Venezuela's inflation rate was a relatively moderate 34.2 percent; Peru's an explosive 409.5 percent. Capital investment levels in Venezuela were around double Peru's ($9.55 billion versus $4.86 billion), while Venezuela's healthy current account surplus compared favorably to Peru's sizeable deficit.

Not surprisingly, social indicators tell a similar story. Peru's rate of infant deaths per thousand is more than twice the Venezuelan rate, sixty-seven as compared to twenty-seven. Venezuela has a more urbanized and literate population than Peru, with only 16 percent of its workforce engaged in agriculture as opposed to 38 percent in Peru.[19]

As the world knows, Venezuela has benefited enormously in the postwar years from its large, exportable endowment of oil. (Peru has some oil too, but not in such quantities.) Is oil, then, the only difference between Venezuela and Peru? If so, will oil alone be enough to propel Venezuela ahead of the depressing picture painted of Peru in the 1990s? Or have other factors intervened to push Venezuela along a faster track to development than Peru, and will these factors continue to grease the path for Venezuela in the next decade?

Lopsided Technology

An examination of technology factors in Venezuela reveals more advantages than in Peru, but a mixed picture overall. The adult literacy rate of 88 percent is certainly above Peru's 79 percent, but not as much as one might expect given Venezuela's economic prosperity of the oil boom years. Spending on education too naturally exceeds Peruvian levels, but again does not approach OECD levels; the government provides some of the worst health education services in Latin America. Indeed, businesses are plagued by shortages of experienced professionals and tradespeople in a number of key sectors including agriculture, general industry, and health. The country still lacks sufficient managerial ability, technical competence, and labor productivity, while the educational system has only turned belatedly toward scientific and technical skills.[20] Rising demand for a wide variety of skills has impelled the government to launch a selective immigration program aimed at importing 50,000 East Europeans over the next five years. Considering that unemployment is still in double digits, this would seem to highlight glaring deficiencies in the educational system.

As with everything else in Venezuela, inadequacies in the educational system ultimately relate to the lopsided nature of its economy. As one novelist and social critic complained recently, "Venezuela is not a country—it's a country glued to an industry."[21] The capital-intensive nature of the oil industry has reduced interest in developing human capital, and diverted resources from nonoil development into the all-consuming petroleum sector. This has limited progress in other areas, resulting in an inadequately skilled work force and the need to import skilled workers.

In this context, the best thing that could have happened to Venezuela in the long run would have been a serious and protracted downturn in oil prices, forcing the government to focus on nonoil development. Unfortunately for Venezuela, oil prices seem to have stabilized albeit at lower levels. While recent expenditures are up, on the whole very little besides rhetoric is likely to be devoted to human capital. President Rafael Caldera, successor to the ill-fated former President Carlos Andrés Pérez (CAP), will almost certainly increase spending on education, thanks to growing social demands and concerns about the future of oil export earnings.

Thus, this will be a stronger area for Venezuela than for Peru, which cannot depend on stable export earnings to finance educational spending. However, the commitment to develop a fundamental and broad-based pool of human capital is lacking. This oversight could prove crippling in the years ahead, as even oil fades in importance and countries that have focused on developing technological and intelligence-based strengths forge ahead. As one observer notes, "the long-range obstacles are constituted not by natural resources but by the underdevelopment of human capabilities,"[22] a damning statement indeed in our model.

An Ethnic Melting Pot

Lower oil revenues during the late 1980s, coupled with a drastic austerity program imposed by the government in early 1989, have produced an increase in social tensions. A substantial portion of the population, perhaps 30 percent, are malnourished and such public services as health and water are collapsing. Economic reforms instituted by Pérez, while holding out the prospect of longer-term stability, initially widened the gap between rich and poor. By the end of 1991 real minimum wages in Caracas had plunged to less than 50 percent of the 1987 level; explosive growth of over 9 percent during 1992 had virtually no trickle-down benefits for the average Venezuelan, reflecting the weakness of an economic model that has produced a "scandalous concentration of wealth."[23] As of mid-1994, some 43 percent of the people were said to be living in "extreme poverty."

Not surprisingly, these problems have resulted in dramatic increases in crime and corruption. The most telling evidence of social tensions was the rioting of February, 1989. Sparked by bus fare increases (part of Pérez's austerity program), the riots killed about 300 and chastened a country that had believed itself to be free of serious social problems. Even during the boom years of 1991 and 1992, sporadic protests continued to occur. Higher oil revenues may alleviate some tensions, but problems lie just below the surface and could erupt again if economic growth falters. Even with respectable macroeconomic growth, serious issues are roiling the populace. Venezuelans want more responsible leadership and social spending, as well as better distribution of growth.

Social tensions, then, are not insignificant despite Venezuela's impressive record of oil-fueled growth. However, there is one crucial difference between Venezuela and Peru in this regard—the relative lack of ethnic or racial rivalries. Few ethnic or racial conflicts exist in Venezuela, and a combination of races can be found at almost all social levels. Immigrants from different parts of the world, particularly from Europe, have generally mixed with the creole and native people. Indeed, fully 70 percent of the population is considered Mestizo as opposed to just 37 percent in Peru, while Amerinds account for only 2 percent versus 45 percent in Peru. The country's multiracial society enjoys an openness of attitude and class mobility that was supported in the past by its oil income. Racial discrimination is not unknown, but is generally not a serious barrier to advancement. (The ruling *Accion Democratica,* or, AD party chose a black ex-mayor as its presidential candidate for the December, 1993 elections, the first black to run for president.) Mestizos have achieved prominence in politics and business, mobility among the social classes is common, and little of the white Hispanic aristocracy that survives in much of Latin America is present here. As a result, the subtle but definite color-based stratification that haunts Peru has not occurred in Venezuela. Without the sharp edge provided by racial discrimination, social tensions are infinitely less threatening than in Peru. This will prove a major advantage for Venezuela in the next decade, although the collapse of its oil-based development strategy may raise new questions about the quiescence of the marginalized population.

Oil: Blessing or Curse?

Any discussion of natural resources and Venezuela must naturally begin and end with oil. Venezuela has the largest oil reserves in Latin America, and has increased its production since the mid-1990 Persian Gulf crisis from 2 million barrels per day to nearly 2.5 million. With proven conventional reserves estimated at 60 billion barrels, the country now holds 6 percent of the world's total reserves.

As a result, oil has become not only the backbone of the Venezuelan economy, but very nearly the entire economy. The national oil company, PDVSA, pays taxes equal to over 80 percent of its operating profit to the government, which relies on petroleum duties for more than 70 percent of total revenues. Indeed, the country owes much of its 1991–92 economic recovery not to brilliant domestic policymaking, but to world oil market developments.

But is oil a blessing or a curse for Venezuela? As we have come to realize, "prosperity is not assured by the discovery of petroleum, even in the age of OPEC, as the Iranians, Nigerians, and Mexicans will testify. So will the Venezuelans."[24] Oil has diverted attention away from development of human capital and nonoil industry. The government's long-term development strategy is based almost solely on oil and other natural resources. This also perpetuates Venezuela's utter and complete vulnerability to events beyond its control in world oil markets. Fluctuating oil prices did begin to focus government attention on diversification toward the late 1980s. Although the economy is still dominated by oil, moderately successful efforts were made to diversify the country's export base. Nonoil exports are growing, but slowly. Iron ore, steel, and bauxite exports are increasing, while coal reserves and production are also significant. The government is actively promoting tourism as a source of foreign revenue, hoping to attract almost 4 million foreign tourists a year by 1997, compared to 700,000 in 1988.

In this context, though, it is worrisome that most of the government's development plans for the future are centered around "mega-projects" between Venezuelan and foreign companies in the oil and minerals sectors. This plan, heavily dominated by oil, virtually ignores export diversification. Stabilizing oil revenues following the 1990 Persian Gulf crisis clearly reduced the urgency of diversification in government planners' eyes. As technological advances help to reduce the world's oil dependence (especially for expensive, heavy crudes like Venezuela's), the country will come to regret its shortsightedness.

Perhaps one hopeful track in the opposite direction is the privatization of the local telecommunications network, CANTV. GTE, the company that purchased CANTV in 1992, is putting close to $4 billion into the local company to upgrade the telephone system. This effort to improve Venezuela's communications infrastructure is critical if the country is to advance economically. It also has a political context. As one Caracas worker commented to the authors in December, 1992: "If the telephone system works, then I'm for the economic reform process."

Sequencing: An Old and Stable Democracy?

Widespread faith in the durability of Venezuela's democratic traditions was shattered by the February and November, 1992 rebellions among military officers, which came perilously close to unseating the government. The military revolts were both puzzling and disturbing, given Venezuela's status as one of Latin America's most venerable and stable democracies. It had been thirty-four years since the last military dictatorship was toppled (in 1958), and the political system had long been dominated by two highly centralized, corporatist (some would say Leninist) parties. However, the country was electrified once more by the suspension of President Pérez in mid-1993 on corruption charges and his replacement by interim President Valasquez.

At first it seemed easy, even natural, to blame Pérez's economic austerity measures for the discontent that fueled the military revolt and eventually brought about his downfall. However, a closer look at the political system reveals that the seeds of discontent are less economic than political. Interestingly, the officers in both coup attempts hoped to eliminate what they viewed as an inefficient and corrupt political system, as well as repeal the government's three-year old economic program. Dozens of violent, antigovernment protests in key cities have highlighted these concerns, revealing a deep disillusionment with the political parties and system itself. The coup attempt struck a sympathetic chord with many Venezuelans, especially among the middle and lower classes. In a way, the surprisingly high level of passive popular support for the rebels' first attempt is reminiscent of Peruvians' support for Fujimori's autogolpe.

Yet, there also was considerable revulsion at the appearance of members of Red Flag, a leftist group that backed the second coup attempt in November, 1992. Members of Red Flag took over the television station and appeared on air, exhorting the people to rise up and take to the streets to overturn the government. Many Venezuelans, who might have supported a coup led by clean-cut military officers promising peace and stability, were disturbed by the appearance of "armed thugs" on television. As one Venezuelan commented, "It was a shock at how close we came to losing something we had taken for granted."

Pérez's ouster was another attack on "democracy" in a sense, indicting the oil-rich but corrupt system in which one-third of the population lives in abject poverty while the politicians are perceived to live like kings. One writer lauds the fact that seven successive presidents have abided by "the rules of the game while guiding economic modernization fueled by vast petroleum and other natural resources."[25] But it is increasingly clear that the "rules of the game" include massive corruption and venality.

If democracy itself is on trial, then the democratic institutions must be called on to justify their legitimacy. In recent years, the two dominant political parties in Venezuela have fallen well short of their democratic responsibilities, to put it mildly. They have been discredited by wide-ranging and well-publicized corruption scandals, increasing public support for emerging left-wing and radical groups in recent elections. As poverty levels have risen in response to austerity, so has anger at high-level corruption soared. Popular hostility to the government and political elite has mounted in response to a series of corruption scandals.

Even so, the military's attempt to destroy Venezuela's much-vaunted democratic stability was shocking. More than economic austerity, the country's legendary corruption is at the root of the problem. There is no doubt that the political system has long tolerated, even encouraged, official corruption. The government party appoints judges who do its bidding, doles out lucrative jobs and contracts, and traditionally receives kickbacks from business. But a generally corrupt legal and judicial system, coupled with rampant corruption among high-level public officials, has become intolerable.

As in Peru, this has resulted in a surprisingly widespread disillusionment with those institutions that purport to be democratic, as well as with democracy itself, and a willingness to support, or at least accept, a return to authoritarianism. This suggests a slightly different set of sequencing problems than in Peru. Democracy without a fundamental reform of the judicial and political systems invites corruption. In turn, it is very difficult to implement economic reform in an atmosphere of blatant, high-level corruption. It is excruciatingly painful to open up an economy to the chill winds of international competition while allowing the country's top officials to continue publicly enriching themselves. The result: a deeply flawed "democratic" system that is having considerable difficulty proceeding with economic reforms.

In this context, the outlook for Venezuela is hardly less bleak than for Peru. Before his ouster Pérez had promised wide-ranging judicial, financial, and constitutional reforms coupled with renewed efforts to punish corruption. Obviously, only moderate results are evident so far, and the highest possible level of skepticism on future results seems appropriate. As high-level corruption saps the energy and credibility of the political elite, substantial reforms will be impossible and social tensions will ferment. The 1993 election of Rafael Caldera, who also previously served as the country's president, boded poorly for Venezuela's prospects, as he has indicated a hazy, populist approach to curing the country of its ills.

Economic "Reform," Venezuelan Style

In line with the rest of the developing world, Pérez was ostensibly committed to a program that aimed to promote a free market economy, involving lower government controls and higher levels of foreign investment. To this end, a sweeping liberalization of the foreign investment code was undertaken. Reforms aimed at winning over foreign investors included cuts in the corporate tax rate, banking reform, and some streamlining of bureaucratic procedures. Most important, the government under Pérez began to actively encourage foreign investment in the oil industry (this by the same president who expropriated private oil companies in 1976).

Moreover, serious austerity measures were implemented between 1989 and 1993 despite difficult economic and social conditions. Reforms initiated in 1989 spurred a sharp decline in economic growth and a painful drop in living standards. Nevertheless, Pérez pressed on with plans to lift many subsidies, reduce import duties, and slowly restructure the public sector through deregulation and privatization. During 1991 the economy grew by over 9 percent and earned $15 billion from oil exports, but Pérez still kept social spending down and permitted real wages to decline. Privatization began slowly for several reasons, including stiff domestic opposition, but finally began to pick up steam in 1992.

In spite of this seeming progress, however, even under Pérez Venezuela did not fundamentally embrace free-market economics and is not likely to do so in the future. Despite half-hearted attempts to diversify the economy, petroleum still accounts for more than 80 percent of foreign exchange earnings and 20 percent of GDP. The budget deficit in 1993 reached 8 percent of GDP, inflation topped 40 percent for the year, and real interest rates were 40 percent or more for commercial loans. Congress was caught up in the storm over Pérez and failed to enact bills aimed at economic reform, including crucial new taxes to help close the fiscal gap. Moreover, major candidates for the 1993 presidential elections attacked Pérez's unpopular free-market policies and promised a more populist stance. The eventual victor, Rafael Caldera, ran on a strongly populist platform. His first year in office witnessed a spate of highly interventionist measures and important retreats from unpopular market reforms. By mid-1994, with the economy contracting and inflation on the rise, one rating agency noted the "disarray in the Caldera administration's economic policies."[26]

Unlike in other Latin American countries, economic reform is still not politically popular with the voters. Probably reflecting at least in part the growing outrage over the unpleasant combination of austerity and corruption, the public is unconvinced of the rectitude of free-market policies—which violate the deeply felt belief that the state should distribute Venezuela's oil wealth. Indeed, nationalist and statist views are still strong, especially in the military. According to one observer, Pérez's downfall can even be interpreted as the "revenge of anti-reform dinosaurs in politics and in the private sector who are opposed to the market-oriented policies Mr. Pérez initiated in 1989."[27]

Against this background, the outlook for meaningful reform is, at best, uninspiring; at worst, grim. Unlike Chile, Mexico, and even Argentina, Venezuelan economic reform is vulnerable to the assault of "dinosaurs." Serious action is likely to be blocked by the "hyperactive legislature," partly reflecting the perception of the general public that broad-based growth is not ahead. The list of interests that brought down Pérez and will continue to delay reform include business magnates accustomed to protection and subsidies; aging dinosaurs and populists within both major parties; and various statists from the openly socialist groupings. Pérez's departure also spells the dislodging of a new generation of reformers from the government, a team of young technocrats who were recruited from the universities and private sector and who were truly committed to economic liberalization.[28] Caldera faces a declining economy, low international oil prices, demands for greater spending, accelerating inflation, and a banking system in crisis. His first move: He suspended constitutional provisions of freedom of economic activity without government restrictions.

This move highlights the most important obstacle to economic reform in Venezuela: the fundamental lack of commitment to free market economics within Venezuelan society and government. The government is still by far Venezuela's most important economic actor. Extensive economic involvement by the government includes its ownership of the petroleum industry; in addition, the state owns or controls a wide range of other heavy industries, service companies, and financial institutions. Even excluding oil, the public sector accounts for 28 percent of GDP. The government actually increased its payroll by 150,000 new bureaucrats during the alleged austerity year of 1990. The constitution is profoundly statist, and it is hard to see how a free market can flourish under central government planning.

The precedent of economic reform set by Venezuela's Latin American neighbors is a powerful influence, but the change in philosophy is not yet politically consolidated. If a country has had a relatively high level of development, it may need to hit rock bottom in order to develop a societal consensus that serious reform is needed. This happened to Mexico in 1982; to Argentina in 1989. Cushioned by its oil wealth, Venezuela has avoided such a plunge. As a result, the country is still fundamentally committed to a statist rather than free-market philosophy. Many future reforms, therefore, are likely to be little more than cosmetic. Limited privatization and stable oil revenues will permit some noninflationary growth in social spending. Reform Venezuelan style, however, is hardly designed to produce a free-market, open economy.

Who Cares?

Does it matter, as long as the oil money keeps rolling in? Although shaken by the abortive rebellion and sudden departure of Pérez, many investors seem to think not. Within hours of the coup attempt, local and foreign investors signed $16 million in contracts to buy four state hotels, while Eastman Kodak inked a letter of intent for a $200 million joint venture. Major investments committed by foreigners as of March, 1992 totalled $24 billion according to the government (although over half of this amount relates to projects not yet begun). Even the arrival of a new populist president in early 1994 has not entirely daunted investors.

But it does matter, nonetheless. First of all, there is the pesky but never-ending problem of volatile world oil prices given the country's excessive dependence on oil revenues. Second, the heavily skewed economy will be unable to generate sufficient employment opportunities or create a substantial pool of human capital to face the challenges of the 1990s. This suggests that Venezuela is really not very different from Peru, except for its oil and lack of ethnic rivalries.

These are still big differences, though, which will help to cushion Venezuela from some of Peru's ills. Oil revenues and relative social peace will allow the country to address its problems somewhat better than Peru. Paradoxically, however, oil revenues will also prevent the country from developing in ways geared to the specific challenges of the 1990s. Venezuela will continue to develop at a higher level and faster pace than Peru, but it is not destined for new-tiger status.

Notes

1. *The Financial Times* (10 April 1992).
2. Ibid.
3. Ibid.
4. Sally Bowen, "Foreign Money Pours Back," *Euromoney* (April 1993).
5. See Felipe Larrain and Jeffrey D. Sachs, "International Financial Relations," in Carlos E. Paredes and Jeffrey D. Sachs, eds., *Peru's Path to Recovery: A Plan for Economic Stabilization and Growth* (Washington, D.C.: Brookings Institute, 1991), 228–52.
6. See Carol Graham, "Peru's APRA Party in Power: Impossible Revolution," *Journal of Interamerican Studies and World Affairs 32,* 3 (Fall 1990).
7. Political Risk Services, *Country Report: Peru* (Syracuse, New York: Political Risk Services, June 1993).
8. For further information on Sendero Luminoso see David Scott Palmer, "Peru, the Drug Business, and Shining Path," *Journal of Interamerican Studies and World Affairs 34,* 3 (Fall 1992); David Scott Palmer, ed., *Shining Path of Peru* (New York: St. Martin's Press, 1992); and Simon Strong, *Shining Path: Terror and Revolution in Peru* (New York: Times Books, 1992).
9. *The New York Times* (2 June 1992).
10. John Crabtree, *Peru Under Garcia: An Opportunity Lost* (Pittsburgh, Pa.: University of Pittsburgh Press, 1992), 11.
11. Peter Klaren, "Peru's Great Divide," *The Wilson Quarterly* (Summer 1990), 24.
12. Klaren, "Peru's Great Divide," 24.
13. *The New York Times* (2 June 1992).
14. *The Financial Times* (14 April 1992).
15. *The New York Times* (23 April 1992).
16. *The Financial Times* (3 June 1992).
17. Crabtree, *Peru Under Garcia,* 3.
18. Ibid., 20.
19. Political Risk Services, *Country Forecasts* (Syracuse, New York: Political Risk Services, June 1993).
20. John D. Martz, "Venezuela, Colombia, and Ecuador," in Jan Knippers Black, ed., *Latin America, Its Problems and its Promise* (Boulder, Colo.: Westview Press, 1991), 434.
21. *The New York Times* (21 January 1992).
22. Martz, "Venezuela, Colombia, and Ecuador," 434.
23. *The New York Times* (21 January 1992).

24. Gary Wynia, *The Politics of Latin American Development, 3rd edition* (Cambridge: Cambridge University Press, 1990), 203.
25. Martz, "Venezuela, Colombia, and Ecuador," 427.
26. Quoted in Scott B. MacDonald and Eric Mendelsohn, "Venezuela: Faces Further Downgrades," *CS First Boston Corporate Strategist* (20 July 1994), 35.
27. M. Delal Baer, "Revenge of the Venezuelan Dinosaurs," *The Wall Street Journal* (18 June 1993).
28. Baer, "Revenge."

References

Baer, M. Delal. "Revenge of the Venezuelan Dinosaurs." *The Wall Street Journal* (18 June 1993).

Black, Jan Knippers, ed. *Latin America, Its Problems and its Promise, 2nd edition.* Boulder, Colo.: Westview Press, 1991.

Bowen, Sally. "Foreign Money Pours Back." *Euromoney* (April 1993).

______. "Fujimori, a Political Prisoner of the Peruvian Armed Forces." *The Financial Times* (16 February 1994).

Crabtree, John. *Peru Under Garcia: An Opportunity Lost.* Pittsburgh, Pa.: University of Pittsburgh Press, 1992.

The Financial Times (10 April 1992).

______. (14 April 1992).

______. (3 June 1992).

Graham, Carol. "Peru's APRA Party in Power: Impossible Revolution." *Journal of Interamerican Studies and World Affairs 32,* 3 (Fall 1990).

Klaren, Peter. "Peru's Great Divide." *The Wilson Quarterly* (Summer 1990).

MacDonald, Scott B. and Eric Mendelsohn. "Venezuela: Faces Further Downgrades." *CS First Boston Corporate Strategist* (20 July 1994).

Nash, Nathaniel. "No Longer a Pariah, Peru is being Recast as Business Magnet." *The New York Times* (2 November 1993).

The New York Times (21 January 1992).

______. (23 April 1992).

______. (2 June 1992).

Paredes, Carlos E. and Jeffrey D. Sachs, eds. *Peru's Path to Recovery: A Plan for Economic Stabilization and Growth.* Washington, D.C.: Brookings Institute, 1991.

Palmer, David Scott. "Peru, the Drug Business, and Shining Path." *Journal of Interamerican Studies and World Affairs 34,* 3 (Fall 1992).

Palmer, David Scott, ed. *Shining Path of Peru.* New York: St. Martin's Press, 1992.

Political Risk Services. *Country Forecasts.* Syracuse, New York: Political Risk Services, June 1993.

______. *Country Report: Peru.* Syracuse, New York: Political Risk Services, June 1993.

Rudolph, James D. *Peru: The Evolution of a Crisis.* Westport, Conn.: Praeger Publishers, 1992.

Strong, Simon. *Shining Path: Terror and Revolution in Peru.* New York: Times Books, 1992.

Wynia, Gary. *The Politics of Latin American Development, 3rd edition.* Cambridge: Cambridge University Press, 1990.

9

India

India has never significantly caught the imagination of American business. While the 1.3 billion people of China fascinate entrepreneurs as a vast, untapped consumer market, India's population of 850 million has never had the same cachet. Although India may take second place in demographic terms, its size far exceeds that of the third-largest country, the United States, at 250 million. In fact, the number of India's inhabitants alone exceeds that of Africa, Europe, or the Americas, making its nickname, the "Subcontinent," seem almost too modest.[1]

If this vast nation can become one of the world's new tigers, it would make the accomplishments of the Pacific Rim nations pale in comparison. India's government has embarked on an ambitious economic reform program designed to stimulate modernization and growth—in effect, to transform India from a lumbering economic elephant to a sleek tiger. The crucial question remains whether there is a national consensus that India should take this new path or continue on a more autarkic, socialist road.

On a more practical level, this program must deal with the country's high poverty level, the intrusion of government bureaucracy into economic life, and the inefficiency of industry. However, its largest challenges may be dealing with political and cultural factors: the tendency to desire autonomy, a distrust of foreign influence and investment, and above all, the societal fissures along ethnic and religious lines.

Historical Baggage

India's experience of living under and throwing off British colonial rule has left it saddled with some historical baggage antithetic to free markets, international trade, and the current thrust in favor of economic reform.

When it became a self-governing dominion within the British Commonwealth in 1947 and fully sovereign under its own constitution three years later, an independent India viewed its past international trade with Great Britain, and especially the British East India Company, as the means for the exploitation of India's people and natural resources for the enrichment of Great Britain and its Empire. The response of India's economic policymakers was to view international trade, the epitome of capitalism, as a threat to Indian strength. Thus, under Jawarharlal Nehru (who succeeded the slain Mahatma Gandhi in 1948), a key economic goal was self-sufficiency, to avoid an over reliance on or vulnerability to other nations, particularly Western Europe and the United States.[2]

In the bilateral cold war framework, the most immediate alternative was to borrow from the Soviet method of planned production and industrialization; the U.S.S.R. was a credible model on the heels of its victory in World War II and the conversion of bucolic Russia into the industrial Soviet state. The resulting system of "command capitalism" provided Nehru with the means to build an economy that provided the products necessary to approach self-sufficiency. A huge official bureaucracy (one remnant of British rule that was embraced) oversaw the central planning process through an increasingly intricate system of licenses, subsidies, and regulations. Forty years of this program left India one of the world's smallest exporters (by share of total output) and with a deeply ingrained attitude of self-sufficiency.[3]

This attitude was a key to India's status as an "elephant" over the four decades since independence. Its 5.4 percent GDP growth rate in the 1980s was an improvement on the 1970s 3.4 percent production growth, but still lagged the 6.0 percent and 4.5 percent growth logged by all low-income countries in those respective decades. Furthermore, with a 2.2 percent population growth rate during that generation, India seemed mired in "have not" status, lower on a per capita income basis than even Haiti, the poorest nation in the Western Hemisphere.[4]

Technology Power

India has been long known for the highly skilled and educated professionals that are the product of its universities and secondary schools. The weakness of India's educational system lies in a lack of breadth rather than depth. The significant improvements in the number of primary and secondary schools has been reflected in more than a doubling of the literacy rate in the three decades ending 1981. Key shortcomings still exist: females are outnumbered three to two in schools and only a third of primary school students finish sixth grade. As a result, India's 36 percent literacy rate falls short of many other developing countries including Indonesia (62 percent) and China (77 percent).[5]

The lack of technological openness is reflected in most of the public enterprises that dominate many industrial and service sectors. Reflecting the country's historical emphasis on self-reliance, these firms were not created to fill profitable niches. Rather, they achieved large size and diversification at the expense of efficiency and competitiveness. While their deficits have not proven to be the major drain on public finance (that honor is due to food subsidies, salaries, fertilizer, and defense), they have not provided a good return on the funds invested in them over the years.

Individual Indian professionals are often exposed to new ideas and technologies as they form a skilled cadre of workers in the United Kingdom, the Middle East, and elsewhere. But India itself has been resistant to these influences; past trade and investment policies focused only on the negative effects to the established government and industrial interests of openness to foreign ideas and products rather than the benefits that would accrue to the entire economy from lower costs, greater competition, and investment capital.

Fortunately for India's economic prospects, this approach is changing through the trade and investment liberalization contained in the government's reform program. This liberalization is in large part a reflection of the power of the growing middle class. While large disparities in income have left 200 million Indians below the poverty line, the size of the middle class has quadrupled over the last twenty-five years to represent 100 million people, or 12 percent of the population. Even the poverty figure is much improved; according to government statistics, in fifteen years the proportion of Indians living in poverty declined to less than 25 percent from 48 percent.[6]

These official figures probably obscure the true progress. Some surveys have put the middle-class figure as high as 350 million. The discrepancy is explained largely by the nationwide tendency for individuals to downplay their personal income in order to avoid joining the 1 percent of the population that pays income tax. This black economy is centered in the middle and upper classes, and could represent 20 to 30 percent of GDP.[7]

With the more open approach to investment and foreign trade, Indian professionals are finding more opportunities to pursue specialties in India. The southern city of Bangalore, dubbed the "Silicon Valley of India," is viewed by economic reformers as a prototype for a modern, competitive, and international India. Bangalore has attracted a number of high-technology investors, especially the large U.S. computer firms, with its moderate climate, its distance from northern religious violence, and, most importantly, its well-educated and inexpensive labor pool. It also provides a guide for further improvement in India's quest for attracting foreign investment: better telecommunications, consistent power generation, adequate water, and international transportation linkages.[8]

Sequencing: Money Before Politics

India is the world's largest democracy, a federal state ruled through a parliamentary system. The weight of power has come to rest with the central government in New Delhi rather than in the twenty-two states and nine territories. In addition to the "normal" power wielded by the center, the federal government has not been loathe to use its authority to unseat local governments and rule states directly from New Delhi when it decides that national security, public order, or its own power base is in jeopardy.

The Congress party has dominated federal politics and it appeared to be the dynastic possession of the Gandhi-Nehru families until a tragic succession of events took place: the assassination of Indira Gandhi by a Hindu extremist in 1984; the earlier death of her eldest son and heir-apparent, Sanjay Gandhi, in a 1980 stunt plane crash; and finally, the assassination of her remaining son, Rajiv Gandhi, in 1991 by a bomb blast while campaigning in southern India. In June, 1991, Prime Minister P. V. Narasimha Rao and his finance minister Manmohan Singh ushered in a new Congress party government dedicated to economic recovery and reform.

One outgrowth of this kind of political domination has been widespread corruption of government officials. The 1993 financial scandal in New Delhi illustrates the public's willingness to believe that all politicians, especially members of the Congress party, are corrupt. This pervasive belief can cause the mere accusation of corruption to weaken a politician; the most recent victim has been Prime Minister Rao himself and his survival will tell much about the prospects for India's transformation.

In 1993, a man charged in connection with the largest bank securities fraud in India's history made the accusation that in 1991, Prime Minister Rao accepted a 10 million *rupee* ($390,000) bribe. Despite the denials of the prime minister, the precarious position of the accuser, and the lack of evidence,[9] a public opinion poll found 43 percent believing the bribery story. Mr. Rao's reputation is also tangentially affected by an investigation into the connection between his son and a $1.6 billion bank fraud.[10]

These charges, which though unproven and probably never to be substantiated, weaken the prime minister in "a country where politicians are held guilty until proven innocent." This can hurt Mr. Rao's push for reform and economic austerity by allowing opponents to the process (such as trade unions, farmers, and other entrenched political groups) to argue to the poor that the sacrifices being called for are not for the benefit of the whole country, but for the narrow interests of the ruling party. In fact, although Mr. Rao need not call an election until 1996, there have been growing calls for him to step down from office coming from opposition parties and newspapers who forced three unsuccessful nonconfidence motions through mid-1993.[11]

Economic Reform

The economic reform package championed by the Rao Government since 1991 is pushing all the right levers. It has ushered in stable macroeconomic policies (with positive results) and a reduction of structural inefficiencies, and built international credibility. India's new attitude and rules toward entrepreneurship and capitalism have attracted foreign investment and reversed capital flight. The main complaint voiced with the plan is that it is too gradual; however, given India's vast size and traditionalism, it is difficult to expect that any significant quickening of pace is politically possible. To the contrary, the striking feature of the program is that so much has been accomplished so quickly.

Indian industry in the early 1990s suffers from years of complacency engendered by protection from both imports and foreign investors who would set up rival firms in India. The first leader to truly address the need for economic reform in recent years was Rajiv Gandhi. Gandhi called for a modernization of industry to be led by the private sector. He began to open Indian firms to a glimmer of domestic and international competition by reducing import tariffs, widening the scope for private investment decisions, and easing licensing rules.

Under Rajiv Gandhi, India looked to have a head start over many economies in that it started with a relatively clean slate: its fiscal deficits, inflation rate, and trade deficits are in manageable limits. (India's fiscal deficit and public sector debt were relatively higher than in many countries with inflationary problems, but India was able to finance its deficits with private savings and foreign concessional lending or aid.)[12]

However, it took an economic crisis caused by reduced foreign aid and workers' remittances, overspending, and a growing reliance on commercial debt to force acceptance of a tough, workable reform program. Events in the early 1990s have kept up the pressure for reform. The Gulf War hurt the economy by continuing to dampen remittances from overseas workers and by bumping up the price of imported petroleum. The dissolution of the Soviet Union upset a major market for Indian goods and a source of subsidized imports (especially energy).[13]

The country had developed a special relationship with the World Bank. In past years, India received as much as 40 percent of the concessionary finance disbursed by with Bank's soft lending arm, the International Development Agency (IDA). However, in the 1980s, the availability of these concessionary loans began to decline under the combined effects of the fiscal constraints on the multilateral development agencies; the increased prominence of the needs of other developing regions, particularly sub-Saharan Africa; and China's entry into the World Bank, which gained it access to a portion of the agency's funding. India's arguments for maintaining its favored position made sense in terms of population statistics, but were weakened by the inefficient use and disappointing results of much of the funds already invested in India's development.

During the same period, India saw reduced flows of foreign exchange from workers' remittances as the Middle Eastern petroleum-fed boom has tapered off. This factor was not as damaging as the concessional debt slowdown because the savings sent home by overseas Indian workers are more stable that those from many other nationalities. One reason is the generally higher skill and educational level of the country's exported workforce. Another is that a large number work outside the states of the Organization of Petroleum Exporting Countries (OPEC), particularly the United Kingdom.

Faced with these declining sources of income, India used its strength to borrow abroad from more expensive commercial sources rather than trim excess spending and move to economic reform. At a time when most sovereign borrowers were either in payment default or had begun tapping the international bond markets, India approached the international banks as a willing borrower with strong foreign exchange and gold reserves, moderate debt service, and a low stock of debt in comparison to size of the economy. International bankers were all too willing to ignore the crucial negative factor that India's debt was high relative to the small and underdeveloped export sector that they were relying on to generate the foreign exchange necessary to repay their loans.

The turn to international financial markets pushed the Indian economy into such an imbalance that a deep-seated economic reform and austerity package became politically saleable. The borrowing directly boosted India's debt service burden and prompted record budget deficits. In 1990–91, high petroleum prices and capital flight combined to drain India's foreign exchange reserves. Facing a default on foreign debt as the alternative, Rao and Singh sold an International Monetary Fund-approved stabilization program to devalue the rupee, control inflation through fiscal austerity, and open up the economy with market-oriented reforms.[14]

The reform package is bearing fruit in purely macroeconomic terms. During the two-year period since Prime Minister Rao entered office in June, 1991, the budget deficit was cut from 8.3 percent to 4.7 percent of GDP projected for 1993. Blessed in part with a good monsoon, inflation came down from the August, 1991 peak of 17 percent to 6 percent. GDP growth has recovered to 4.2 percent in 1992–93; while this is below the rate the World Bank considers necessary for sustained improvement in national welfare, it is a strong improvement from 1.2 percent in 1991–92.[15]

The devaluation of the rupee was the first step in efforts to stimulate exports. This was especially critical after the collapse of the old Soviet bloc left India bereft of its former trading partners and contributed to a disappointing 3.6 percent pace of export growth in 1992. Another step was the decision to make the rupee convertible for the settlement of trade accounts; the impact of this move was felt immediately with a 29 percent export jump in April, 1993 and an unanticipated trade surplus. The trading sector was liberalized in nonfinancial terms as well: the maximum import tariff was reduced to 85 percent from 200 percent, and import quotas abolished for all but consumer goods.[16]

Similarly bold steps were taken to open up the general economy. In 1993, industrial licenses were required for only fifteen industries, and the public sector's monopoly has been pruned to six industries from seventeen. Ownership stakes of up to 20 percent have been sold in these public companies, and their subsidies have been curtailed. Foreign institutions are allowed to invest in India's stock markets, whose attractiveness is enhanced by the use of British law and accounting standards. This package helped bring in proposals for $2.2 billion in foreign investment: half of China's 1991 total, but almost fifteen times the average India attracted in the 1980s. For the first time, in 1993 a private-sector Indian firm, the Steel Authority of India, was included in the *Forbes* magazine list of the 500 largest public foreign (i.e., non-U.S.) companies at rank 488.[17]

For India, the move to welcome foreign investment is a true sea change given the national scars from its colonial experience; India "has finally decided to stop fighting the East India Company."[18] Additionally, India's more open economy and its greater emphasis on exports has helped build up a comfortable foreign exchange reserve level, from the 1990's low of $1.5 billion to $20 billion in early 1995.[19]

Agriculture, Energy, and Natural Resources

Three central resource issues for India are agriculture, energy, and population. The implications for India's transformation are mixed. On balance, agriculture has been a success story. India has shifted from the danger of periodic famine to a food surplus. This remarkable achievement has come at a high cost in terms of the diversion of development funds, and India has ended its reliance on the annual monsoon. However, energy and population growth remain important challenges. Both will keep the pressure on economic performance to expand exports and to provide a widening employment and income base.

Agriculture

The agricultural sector is critical in India. While it accounts for 31 percent of GDP, it accounts for 70 percent of employment and 80 percent of the poor live in rural areas. While its production growth has lagged behind that of other sectors and consequently reduced its share of national output, the $78 billion in agricultural value added is exceeded only by China. Besides meeting the nation's food needs, the agricultural sector provides the raw materials, such as jute and cotton, for much of domestic industry.[20]

This sector has been the focus of intensive development, largely supported by foreign aid. The primary focus has been on boosting foodgrain production through the introduction of high-yield varieties of rice and wheat, heavy use of fertilizers, and reliance on irrigation. In normal years, this combination of efforts dubbed the "Green Revolution" created an overall surplus of food production over consumption. This surplus is fed into a nationwide system of food stockpiling and distribution. When a monsoon failure occurs (every five years, on average), the government is able to disseminate food reserves to prevent famine.[21]

India has boosted per capita food production by 19 percent from 1979/81 to 1988/90. Nonetheless, the unchanged reliance on the yearly monsoon is behind the country's wide swings in agricultural production; in turn, this means that economic performance remains vulnerable to the success or failure of the monsoon. The 1993 monsoon season provided a particularly deadly reminder of this vulnerability: the July monsoon floods left 3,000 dead and destroyed the homes of millions across Northern India, Bangladesh, Nepal, and Pakistan.[22]

Energy

India is also vulnerable to external changes in the international petroleum market. While its workers benefitted from wages earned in booming OPEC countries, these gains paled in comparison to the increased cost of imported hydrocarbons. India has gotten positive results from its efforts to boost its domestic oil production, but the potential for increasing gains seems limited without the discovery of major new reserves. Production of natural gas is responding to favorable government policies, but this resource is unlikely to become a critical component of the energy mix because of the intensive work needed to develop a market for gas and to construct the delivery and processing infrastructure. Fuels account for one-sixth of merchandise imports (up from 5 percent in 1965), leaving India's trade performance vulnerable to energy price swings.[23]

Population

India's population is expected to reach 1 billion (nearly one-sixth of the world total) by the end of the century, even if official projections of slowing growth prove correct. This demographic expansion will continue to put pressure on the pace of economic growth in general and employment and agriculture in particular. With a population growth rate of over 2 percent, the World Bank has projected that the economy must grow at least 5 percent annually on average to allow for better living standards, employ new entrants to the work force, and reduce poverty. The required growth rate exceeds the demographic rate partly because the labor force will be increasing more rapidly than the overall population because 40 percent of Indians are under fifteen years old.[24]

Government-led efforts to moderate population growth have met with only partial success due to resistant cultural attitudes and to a backlash against excessive pressures for people to undergo sterilization. In fact, improvements in the delivery of health services have the negative side effect of further boosting the growth rate. Generally, parents do not limit their families to one or two children; the preference for larger families is fueled in part by the perceived advantage of having sons.[25]

Growth will be felt increasingly in urban areas due to immigration from rural regions and natural increase. The population in the cities, about 156 million (24 percent of total) in 1981, is almost certain to more than double by the year 2000. With infrastructure already inadequate in many cities, India will face steep challenges in ensuring adequate water, sanitation, transportation, and housing.[26]

Pressure from the size and growth momentum of India's population is one side of the issue. The other is the sheer diversity of India's peoples from a number of perspectives. Geographically, India covers a territory almost as large as Europe containing a range of ecosystems from jungle to desert, mountains to seashore. Linguistically the country is home to fifteen major languages and some 874 regional dialects. The major religious fault is between the 85 percent of the population who are Hindu and the Moslem 10 percent. Nonetheless, other faiths command the loyalty of sizeable flocks: Christianity at 20 million adherents, Sikhism at 14 million, Buddhism at 5 million, and Jainism at 4 million.[27] These differences can be a source of Indian strength and pride, but also underlie the often violent divisions within society and resistance to national authority.

Political Centralization

The key to determining the potential for India's transformation from elephant to a tiger is the ability of the central government to implement its reform efforts or, conversely, the willingness of the nation to accept the direction of the center. This reform effort is not simply a choice of tactics; rather, it is a fundamental strategic shift that will need the stated or tacit acceptance of the nation to succeed. Is there a consensus on a move from a socialist to a capitalist economy? Is there a consensus on whether the public sector should be profitable?

A negative answer to those questions will arise in one of two incarnations. The elites are represented by the bureaucracy, which challenges the central authority through its ability to stifle change. The mass of people are split along religious, ethnic, and caste lines; these are the fissures along which they will threaten nation unity if they fundamentally disapprove of central authority. If Rao's reform program survives the intrigues of the bureaucracy and the unrest of the masses, the transformation will succeed.

Rao's reform program has produced rapid results, which are even more impressive when compared with the results most countries have achieved with IMF stabilization policies. The reluctance of debt-rating agencies to restore India to investment grade status, however, reflects the underlying political uncertainties.[28]

One challenge to the centralization of power and to Rao's program lies within the government itself. Perhaps the most enduring legacy from the British raj, India's bureaucracy is justifiably famous for its ability to mire the economy in red tape. Despite the positive effect of the Rao reforms on its upper levels, bureaucratic delays and obstructions remain daunting at lower levels of the federal government and in the state governments. For example, the reforms make the building of a factory much easier, but that factory cannot be closed or cut back without the express permission of the state government. Since that permission is rarely forthcoming, companies often find that their only economic alternative is to abandon a loss-making facility.[29]

The explicit rules of the bureaucracy are only part of the problem. The other is the rampant use of official positions—beginning at the lowest levels—to extract bribes for favorable decisions or simply to provide basic services. The streamlining of government regulations has reduced the potential for graft, but middle and lower level officials are still able to extract payoffs. While the bureaucracy provides the means, the underlying cause is twofold: the widespread poverty, which makes the recipients rely on graft for subsistence; and the shortage of goods and services, which presents the consumer with the stark option of paying the bribe or going without the product.[30] In the words of one commentator writing in 1992, "Today, it is impossible to get anything done in India without bribery or commissions."[31]

The more immediate and more dangerous challenge is the potential for the violent fracturing of India itself. In the past, those wishing India well counted on its sheer physical size, its immense population, and the diversity of its peoples, religions, and languages to dilute or absorb the danger of specific rivalries. This diversity frustrated those attempting to affect true centralization of political power, but it also prevented the nation from coming apart at the seams.

In this context, the recent rise of the Bharatiya Janata party (BJP) is particularly worrisome. While its political platform positions the BJP as a free market and uncorrupted alternative to the Congress party, the real source of its strength is its unabashed pro-Hindu stance. Since independence from the British empire, the Indian government (i.e., the Congress party) has tried to rule India in a nonsectarian manner. The BJP has chosen to take the opposite tack by exploiting the perceived grievances of the Hindu majority against the Moslem minority. It claims that the Congress party and other nonreligious parties curry favor with the Moslems to buy their votes, and it even complains about ancient history in the form of the Islamic invasion of India.[32]

Its support has never gone over 20 percent of the vote, but it appears to be gaining support from a larger portion of the Hindu community with each major political event: the latest Congress party corruption scandal, riots surrounding the Hindu razing of a mosque, and the apparently antidemocratic suspension of four state governments by the federal government in the wake of the riots.[33]

Prime Minister Rao hopes to hold the Islamic vote by not acceding to the BJP agenda. But the strategy faces a potential threat from the Janata party, which used to be Congress's main opposition and which remains a viable force in northern India. Depending on Janata's ability to solve its internal divisions, it could attract the Moslem vote in Uttar Pradesh and push Congress into third place in the nation's largest state.[34]

Some small consolation can be taken in the BJP's profession that it is more pro-reform than the Congress party and in the probable strategic response of the Rao government. With the opposition taking away the religious card, it must concentrate on delivering superior economic performance to convince the electorate through its wallets. Hysteria on the religious front distracts attention from the pain during the time it takes the austerity program to yield results.[35]

Unfortunately, this is small consolation indeed, especially as without a booming recovery at the next election, the more likely scenario is for the voters to blame the Rao administration for the austerity rather than his predecessors who created the need for it. While much of the BJP's attraction can be traced to economic grievances, the mere sharing of India's economic and political success does not seem to inoculate various group's from violence and rebellion.[36]

Political Flashpoints

The current incidents of violence, rebellion, and civil war are all the more worrisome if the BJP proves able to harness Hindu nationalism into a political force. With that prospect in mind, observers of Indian politics are closely watching three arenas in particular: Uttar Pradesh in the wake of the Ayodhya riots, the civil war in the border state of Kashmir, and the recently quieted Sikh rebellion in the Punjab.

Uttar Pradesh. On 6 December 1992, a sixteenth-century Moslem mosque in the northern city of Ayodhya in the state of Uttar Pradesh was destroyed by a 200,000-member Hindu mob. The resulting riot left 1,190 dead according to (admittedly conservative) official statistics. The large majority of these were killed by the police, who are accused of targeting Moslems. Unlike most past violence, the rioting did not seem to act as its usual escape valve and religious tension remains unchecked.[37]

The Moghul emperor Babur built the 400-year old Babri mosque on what many Hindus believe to be the holy birth place of the god Ram. More worrisome is the ancillary claim that the mosque also stood on the site of a Hindu temple that was razed by the Moslem invaders to erect their own place of worship. This justification for pulling down mosques heralds future problems since a number of them were built by the Islamic invaders on the sites of Hindu temples. Hindu militants would raze all the mosques built by the Islamic invaders.[38]

Faced with growing Hindu sentiment against the offending mosque, the federal government and more moderate BJP leaders were willing to debate (and allow to drag out) the fate of the Babri mosque in court, but the BJP was pressured to hold a rally on the site. While the leaders of both parties planned on a peaceful demonstration, it soon went out of control with the crowd tearing down the mosque and erecting a makeshift temple to Ram in the rubble.[39]

In the wake of the riot and the involvement by local BJP officials, New Delhi banned several militant religious groups such as the Hindu Rashtriya Swayamsevak Sangh (the National Volunteer Corps, or RSS) and the Islamic Sevak Sangh. As the RSS's leadership forms the core of the BJP, New Delhi also dissolved the BJP-ruled state governments covering most of India's Hindu heartland—Uttar Pradesh, Rajasthan, Himachal Pradesh, and Madhya Pradesh—and representing one-third of India's population, in distrust of the BJP administrations' political willingness to enforce the RSS ban.[40]

Despite New Delhi's additional step of arresting some 5,000 high- and middle-level leaders of the RSS after the December, 1992 riot, the Corps reportedly continued to function relatively unimpeded. Perhaps this was due to RSS's experience in surviving official censure; it was banned twice before: in 1948–49, after Mahatma Gandhi was assassinated by an RSS member, and in 1975–77, during the emergency rule of Indira Gandhi. A key to its survival is about 5 million former RSS volunteers, many of whom have become highly placed in Indian government, industry, and society over the years. In a partial reflection of the strength of Hindu nationalism and the BJP's lengthening political shadow, Uttar Pradesh's High Court reversed the prohibition on the RSS at the same time as another court was confirming the ban on the Islamic Sevak Sangh.[41]

In the aftermath of the riot, Rao initially pledged to rebuild the mosque and build a Ram temple nearby. But support is running high in Uttar Pradesh for permanently replacing the mosque with a Hindu temple. Rao has temporized in the hopes that the issue can be settled in the courts and in the hands of religious leaders.[42]

The Hindu fanaticism may be breeding an Islamic backlash. The civil wars in Punjab and Kashmir already have a religious flavor as Pakistani Moslems support the rebels in spirit and with weapons. While Rao hopes for the passage of time to cool tempers, the outnumbered and outgunned Moslem minority may turn to terrorism to exact its revenge. This appears to be the underlying cause of the riots in January, 1993 and bombings in March, 1993 in, of all places, Bombay. The business heart of India, this bustling city was thought to be immune from religious strife as its 12 million people seem to exist only for the single-minded pursuit of making money.[43]

The Punjab. The showpiece of the agricultural "Green Revolution" is the northern state of Punjab. By all accounts, the Punjab should be one of the most contented regions given the resources devoted to transforming it into India's bread basket and the disproportionately large role of the state's Sikh majority in the nation's military, industries, and government.

On the contrary, the Punjab serves as a demonstration of India's vulnerability to religious and ethnic strife despite relative economic success. A violent Sikh independence movement spawned a decade of terrorism that resulted in 25,000 deaths. In 1984, Sikh extremists assassinated Indira Gandhi following the federal siege and her orders to storm the holy Golden Temple at Amritsar where Sikh militants were holed up. While not acceding to demands for an autonomous Sikh state in India, Rajiv Gandhi responded with an accommodating policy by creating a gerrymandered Sikh political majority in Punjab, lifting of emergency military powers, and according greater leniency to Sikh political prisoners.[44]

As late as 1992, government control was limited to the cities in the daytime; at other times and places, Sikh extremists held sway and effectively enforced their austere religious decrees using the stick of about 200 deaths a month.[45] By mid-1993, the threat that Sikh violence posed to the Punjabi people and to the cohesion of the Indian nation appeared past. New Delhi had ruled the state directly for five years; when this rule was lifted, Sikh death threats reduced the turnout for the 1992 elections for the state assembly to 22 percent. In contrast, voter participation in 1993's municipal and village elections was three to four times higher as Sikh threats fell on deaf ears.[46]

The key to this turnaround was the attitude of the new chief minister, Beant Singh of the Congress party, who won the low-turnout 1992 election. Bolstered by a steady increase in police strength and convinced that no accommodation with the militants was possible, he gave the police free reign to stop the terror. The police, who had become Sikh targets, turned to torture and killing to suppress the militants. With newfound confidence in police power and tired of the unending violence, the citizenship turned to the police as the best means of ending the war.[47]

The lesson learned from the Sikhs was a troubling one for a democratic state. Compromise and accommodation did not work; the suppression of the revolt by all available force appeared to be the answer. This message was not lost on New Delhi when it turned to deal with the problems in Kashmir.

Kashmir. A bloody rebellion is taking place in Kashmir, in the northern tip of India. It has been a low-profile conflict because events are obscured or complicated by the region's remoteness, the proximity of Pakistan, and the unresolved border issues. This Moslem-dominated state has been the focus of two of the three wars between India and Pakistan, and is considered to be one the world's key flash points, lying between three (with China) presumed nuclear powers.[48]

Long-standing Kashmiri hopes for independence from New Delhi flared into violence in 1989 as rebels, tired of political stalemate, were encouraged by the success of Afghan rebels against Soviet troops. Moreover, they were provided with armaments and other support via Pakistan. The tactics of the roughly 30 militant groups include murder, rape, and extortion, mainly against the Hindu minority.[49]

The response by Indian army and security forces to Kashmiri militancy is reportedly even more brutal. Local citizens have estimated that 12,000 to 20,000 have died in the three years leading to mid-1993, and there also are widespread accusations of government torture and terror. Based on the apparent success in restoring calm to the Punjab, the official strategy appears to be one of massive retaliation. As one example, in May, 1993 government forces torched over 200 houses and stores in response to a rebel attack on an unoccupied military building.[50] Problems have continued in 1994 and 1995.

The Kashmir civil war's impact on India has been compared to that of Vietnam and Afghanistan for the U.S. and Soviet Union, respectively; despite the high human and financial cost of the conflict, there is little prospect of victory or ultimate resolution. In the meantime, the Kashmiri people and economy suffer from the violence, a breakdown in civil order and justice, and the destruction of the all-important tourist industry.[51]

The civil war in the Kashmir is a tragedy for its many victims. However, as with India's other conflicts and unrest, it has a wider impact on the entire population. The high cost of maintaining a military occupation or even simply enhanced police readiness detracts attention and resources from India's overall development objectives. Budget discipline is all the more difficult when the higher military spending is coupled with reduced tax revenues due to lower production in the regions of conflict. Finally, the financing of deficits becomes more expensive when political violence encourages capital flight and discourages foreign investment or Indian remittances from overseas.

Conclusion

India's size, diversity, and traditionalism present some of the greatest challenges faced in transforming an economy. Long proud of its insular, socialist stance, India is being rudely pushed into a competitive capitalist mold by its federal government. Prime Minister Rao was faced with little choice upon taking office other than an aggressive reform program. With the rising threat of Hindu nationalism, he now has little option but to succeed or face electoral defeat by 1996.

The Congress party's reform program has all the ingredients for launching India from the realm of elephants to tigers. Like the proverbial supertanker, this vast nation will turn more slowly than smaller, more uniform states, but the impact of its conversion will be enormous. As one analyst noted in March, 1994: "Many people recognize the difficulties India faces—and the need for further radical reforms. What is missing is any widespread sense of urgency. India, it seems, will modernize—but at its own steady pace."[52] Looking into the next century, India will continue to face serious challenges from its growing population, who will demand jobs, education, and opportunity, from the agricultural sector to keep pace with the demographic demands, and from changes in international energy markets.

Defeat at the hands of the BJP would shake the core of the Indian state. By replacing its secular tradition with an abashedly pro-Hindu one, the BJP would earmark the 170 million non-Hindus as second-class citizens. Increased domestic violence and terrorism by a disenfranchised minority would appear likely, and heightened international tensions, especially with Islamic Pakistan, would probably follow. All of these events would have negative economic impacts and rightfully shake the confidence of international investors.

In the long term, the reformist approach should move India toward a solution of its internal fissures. Old animosities may fade when it becomes clear that personal success depends on a person's efforts rather than their religion, ethnicity, caste, or political connections. In the shorter term, the world will be watching India for signs that the nation will choose to move forward with liberalization, rather than backwards into Hindu nationalism.

Notes

1. World Bank, *World Development Report 1992, Development and the Environment* (New York: Oxford University Press, 1992), 218–19.
2. Eric Pan, "India: The Brightest Jewel Needs Some Polishing," *Harvard International Review* (Fall 1991): 51.
3. Ibid.: 51–53.
4. World Bank, *World Development Report, 1993, Investing in Health* (New York: Oxford University Press, 1993), 238, 240.
5. James A. Hanson, *India: Recent Developments and Medium-Term Issues. (A World Bank Country Study)* (Washington, D.C.: World Bank, 1989), 45.
6. V. G. Kulkarni, "Marketing: The Middle Class Bulge," *Far Eastern Economic Review* (14 January 1993): 44–46.
7. Ibid.
8. John Ward Anderson, "Indians, Foreigners Build Silicon Valley in Bangalore," *The Washington Post* (1 August 1993): A21.
9. It was not only the lack of evidence that had many doubting the charges put forward by Bombay broker Harshad Mehta. Some journalists found the proffered evidence hard to believe. Mehta claimed to have given Rao over 10 million rupees in two suitcases. The first suitcase, allegedly delivered 4 November 1991, was to contain two-thirds of the payment.

   However, upon weighing the largest (500 rupee) bill, the *Economic Times* calculated that the first suitcase would weigh 123 pounds. The *Hindustan Times* refined that measurement to 176 pounds, based on Mehta's testimony that he used smaller bills after being unable to assemble the rarer 500 rupee notes.

   Suman Dubey, "Indian Press Treats Mehta's Allegations as a Weighty Issue," *The Asian Wall Street Journal* (28 June 1993): 10.
10. "India: The Rao Row," *The Economist* (3 July 1993): 32. Molly Moore and John Ward Anderson, "Chronic Indian Bribery Haunts Premier," *Washington Post* (4 July 1993): A22.
11. Ibid. Molly Moore, "Indian Premier Defeats Drive to Oust Him," *The Washington Post* (29 July 1993): A18.
12. Hanson, *India,* 11.
13. Kulkarni, "The Middle Class Bulge": 45.
14. "India: Look Out Asia," *The Economist* (26 June 1993): 34–35.
15. Ibid.
16. Ibid.
17. Ibid. Robert Lenzner, "Bearish on America," *Forbes* (19 July 1993): 104. Eric S. Hardy and Steve Kichen, eds., "The Forbes Foreign Rankings," *Forbes* (19 July 1993): 126–29 ff.
18. "India: Trade without the Flag," *The Economist* (8 February 1992): 34.
19. International Monetary Fund, *International Financial Statistics March 1995* (Washington, D.C.: IMF, March 1995): 289.

20. Hanson, *India,* 25. *World Development Report 1992,* 220, 222, and 224–25.
21. Ibid., 27–28.
22. Ibid., 25. Molly Moore, "Monsoon Floods Spread Destruction Across South Asia, Killing 3,000," *The Washington Post* (31 July 1993): A18.
23. *World Development Report 1992,* 246. Hanson, *India,* 41–42.
24. Ibid., xv, 19, 50. World Bank, *World Development Report, 1993,* 288–89.
25. Hanson, *India,* 19.
26. Ibid., 21–22.
27. Pranay Gupte, *Mother India: A Political Biography of Indira Gandhi* (New York: Charles Scribner's Sons, 1992), 26–27.
28. "India: Look Out Asia," 34–35.
29. Ibid. Moore and Anderson, "Chronic Indian Bribery Haunts Premier": A22.
30. Ibid.
31. Gupte, *Mother India,* 21.
32. "India: To Smile, and Smile," *The Economist* (8 May 1993): 42. "The Fire of India's Religions," *The Economist* (16 January 1993): 33–34.
33. "India: Hindu v. Hindu," *The Economist* (19 December 1992): 35.
34. "India: Holy Help," *The Economist* (12 June 1993): 45.
35. "India: Look Out, Asia": 34–35.
36. Ibid.
37. Cameron Barr, "Indian Report Sees Police Role in Violence," *Christian Science Monitor* (29 December 1992): 6.
38. "Testing India's Structure," *The Economist* (12 December 1992): 41–42.
39. Ibid.
40. "India: Holy Help": p. 45. Hamish McDonald, "India: A Bag Full of Rupees," *Far Eastern Economic Review* (1 July 1993): 15
41. "India: Holy Help": 45. Hamish McDonald, "India: Ayodhya Backlash," *Far Eastern Economic Review* (14 January 1993): 22.
42. "India: Holy Help": 45.
43. "The Fire of India's Religions": 33–34.
44. "India: Peace at Last in Punjab," *The Economist* (22 May 1993): 45.
45. Ibid.
46. Ibid.
47. Ibid.
48. Molly Moore and John Ward Anderson, "Kashmir's Brutal and Unpublicized War," *Washington Post* (7 June 1993): A1, A16.
49. Ibid.
50. Ibid.
51. Ibid.
52. Stefan Wagstyl, "Investment in India, A Survey," *Financial Times (London)* (30 March 1994): I.

References

Books

Gupte, Pranay. *Mother India: A Political Biography of Indira Gandhi.* New York: Charles Scribner's Sons, 1992.

Hanson, James A. *India: Recent Developments and Medium-Term Issues (A World Bank Country Study).* Washington, D.C.: World Bank, 1989.

World Bank. *World Development Report 1992, Development and the Environment.* New York: Oxford University Press, 1992.

World Bank, *World Development Report 1993, Investing in Health.* New York: Oxford University Press, 1993.

Periodicals

Anderson, John Ward. "Indians, Foreigners Build Silicon Valley in Bangalore." *The Washington Post* (1 August 1993): A21.

Barr, Cameron. "Indian Report Sees Police Role in Violence." *Christian Science Monitor* (29 December 1992): 6.

Dubey, Suman. "Indian Press Treats Mehta's Allegations as a Weighty Issue." *The Asian Wall Street Journal* (28 June 1993): 10.

"The Fire of India's Religions." *The Economist* (16 January 1993): 33–34.

Hardy, Eric S. and Steve Kichen, eds. "The Forbes Foreign Rankings." *Forbes* (19 July 1993): 126–29 ff.

"India: Hindu v. Hindu." *The Economist* (19 December 1992): 35.

"India: Holy Help." *The Economist* (12 June 1993): 45.

"India: Look Out Asia." *The Economist* (26 June 1993): 34–35.

"India: Peace at Last in Punjab." *The Economist* (22 May 1993): 45.

"India: The Rao Row." *The Economist* (3 July 1993): 32.

"India: To Smile, and Smile." *The Economist* (8 May 1993): 42.

"India: Trade without the Flag." *The Economist* (8 February 1992): 34.

Kulkarni, V. G. "Marketing: The Middle Class Bulge." *Far Eastern Economic Review* (14 January 1993): 44-46.

Lenzner, Robert. "Bearish on America." *Forbes* (19 July 1993): 102–105.

McDonald, Hamish. "India: A Bag Full of Rupees." *Far Eastern Economic Review* (1 July 1993): 15

______. "India: Ayodhya Backlash." *Far Eastern Economic Review* (14 January 1993): 22.

Moore, Molly. "Indian Premier Defeats Drive to Oust Him." *The Washington Post* (29 July 1993): A18.

______. "Monsoon Floods Spread Destruction Across South Asia, Killing 3,000." *The Washington Post* (31 July 1993): A18.

Moore, Molly, and John Ward Anderson. "Chronic Indian Bribery Haunts Premier." *The Washington Post* (4 July 1993): A22.

______. "Kashmir's Brutal and Unpublicized War." *The Washington Post* (7 June 1993): A1, A16.

Pan, Eric. "India: The Brightest Jewel Needs Some Polishing." *Harvard International Review* (Fall 1991): 51–53.

"Testing India's Structure." *The Economist* (12 December 1992): 41–42.

Wagstyl, Stefan. "Investment in India, A Survey." *Financial Times (London)* (30 March 1994): I.

10

Nigeria

The question of whether sub-Saharan or black Africa can succeed in the new international economy is often distilled into the question: "Can Nigeria make it?" In the wake of the 1980s, Africa's disastrous "lost decade," many pose the query in a more fatalistic form: "If Nigeria cannot make it, how can the rest of Africa hope to succeed?"

In many ways, among the variety of over fifty mainland and island states in Africa, Nigeria represents the best hopes for the region. The most populous country is endowed with enormous oil and natural gas resources that make it an important member of OPEC. A positive legacy of its tenure as a British colony was that the country was left with relatively advanced infrastructure and governmental institutions befitting the centerpiece of London's West African empire. National unity was given a test by fire in the Biafran civil war. It has a diversified demographic, agricultural, and industrial base. As a result of these factors, Nigeria is the undisputed "regional superpower" of West Africa.

The introduction to a survey of Nigeria in *The Economist* summed up Nigeria's importance and disappointment in the eyes of the world, Africa, and itself:

> One in every five Africans south of the Sahara is a Nigerian; the nation is the tenth most populous in the world. Nigeria pumps out 2 [million] barrels of oil a day, making it the fourth biggest producer in OPEC. Its proven natural gas supply could supply Western Europe for at least ten years. Its prickly, proud people number among them the best educated on the continent. Thus Nigerians are forever being told, and forever telling visitors, that they are the giants of Africa. If Africa is ever going to produce a South Korea, they say, it will happen in Nigeria. Yet each time the country has the chance to turn itself into a prosperous model for still-poor Africa, it blows it.[1]

The key for Nigeria to succeed—or become a tiger rather than an elephant—is for democracy to succeed. Many countries look to the successful Asian rim countries as examples of the benefits of the strategy gradual broadening of democratic freedoms following the implementation of central economic policies.

However, Nigeria has lost the luxury of gradualism in politics: Despite the ascendancy of centralist military rule for most of its independent life, the Nigerian government has failed to take the key steps necessary to stabilize its economy and garner the benefits of its natural and human resources. The widespread corruption and waste has fatally weakened the military government's mandate to voluntarily extract the necessary economic sacrifices from its citizens. In an increasingly populist climate, any price hike or tax reform is viewed as a "sacrifice" that will flow into the pockets of Nigeria's leaders rather than accrue to the good of society at large.

Only a truly representative and responsive government can regain the confidence of the Nigerian people; enact the necessary economic reforms; and begin to heal the ethnic, geographic, and religious rifts that are increasingly dividing Nigeria. The 1993 presidential election is examined in some depth to highlight the difficulties of the most serious effort to return Nigeria to popular rule in a decade. That attempt ultimately failed. However, even if a fresh effort ultimately results in the installation of a democratically elected government, the key variable to observe will be whether the military can resist its reflexive tendency to manipulate or depose the civilian authorities.

Resistance to Democracy

Nigeria exhibits many of the political characteristics common to most African countries. First, it is among the newest independent nations in the world and has the most recent memories of colonial status. While there are some exceptions—such as Ethiopia, Liberia, and South Africa, which were already at least nominally independent—almost all African nations gained independence from European rule rather recently.[2]

The transition to African independence began when the British colony of Gold Coast became the new nation of Ghana in 1957. For all intents and purposes, the period of European rule closed in the 1970s with the end of Portuguese rule in Angola, Mozambique, and Guinea-Bissau. However, borders, and indeed national identities, are not set in stone; for instance, 1993 saw South Africa's return of its Walvis Bay enclave to Namibia.

In contrast to the newness of most African nations, the majority of Asian and North African nations gained independence from western rule by the close of the 1950s. Most Latin American countries broke away from Spain and Portugal in the first half of the nineteenth century.

While nations such as France and Great Britain brought indigenous African leaders into the process of colonial government to some degree, few of these new countries possessed an adequate depth and breadth of political leadership at independence. The impact of this inexperience was exacerbated by the problems that most of these fledgling nations faced: border disputes brought about by boundaries that often did not reflect tribal, religious, linguistic, or geographic realities (even when the maps were well defined); infrastructures (in cases where they were not deliberately destroyed, such as by the Portuguese in Angola) that reflected colonial rather than national structures; and a low level of economic development and welfare.

This lack of experience in self-governance by the population and their would-be leaders goes a long way toward explaining the general lack of true western-style democratic government in the continent. Even though the major colonial powers of Great Britain and France had long-established, multiparty parliamentary traditions, most African nations chose a more autocratic form of government. Often this was imposed via a dictatorship (supported by the military, often the sector with the most organizational training and experience), but when the franchise was extended there usually was only one viable political party; thus compromises are made within cliques or parties rather than by popular decision. These political systems increase the opportunities for corruption and cronyism (that goes beyond tribal loyalties) that have become endemic in several countries.

History of Military Rule

Since independence in 1960, Nigeria has experienced only about a decade of civilian rule, but all of five military coups, four coup attempts, and six military rulers. Politics begins with the premise that the military is distrustful of the capabilities and integrity of civilian government. The two civilian administrations' misrule, corruption, and incompetence (especially the most recent period of 1979–83) has underlain this distrust. Unfortunately, despite hopeful beginnings, the results of the military regimes have not presented much of an improvement, although they have ruled for the bulk of Nigeria's first thirty-three years of independence. The contortions of the government of General Ibrahim Babangida to negate the results of the most restrictive of presidential elections added to the perception of self-dealing and incompetence of the military government. Further, it eroded Nigeria's international support, which was predicated to a large extent on buying into Babangida's promises of a gradual return to democracy.

Babangida took power from fellow general Muhammadu Buhari in a 1985 coup while promising to return the country to democracy after an interim period. (Babangida was a key supporter of Buhari's coup in 1983, which ousted the civilian government of Shehu Shagari.) He promised presidential elections in 1990, but that period was stretched out into mid-1993. Ambivalence to a return to civilian rule is not surprising. Along with all Nigerians, the military experienced the failures of the civilian governments. But the armed forces may have the additional concern that they will suffer from civilian-led investigations into government corruption, particularly the widely assumed diversion of oil revenues.[3]

Democratization from Above

Nigeria's new ruler felt that tribal and religious affiliations drove previous elections and civilian governments at the expense of clean and workable government. To avoid these influences, he banned all former politicians from senior office, and created two political parties from whole cloth. The party manifestos placed both the Social Democratic party (SDP) and the National Republican Convention (NRC) in the political mainstream with one slightly more "liberal" and the other more "conservative." He introduced elections gradually: governors and state assemblies in late 1991, and the bicameral National Assembly (formed in anticipation of a civilian president) in July, 1992.[4]

The ironic, but foreseeable, result of this ideological sameness was that the parties turned to tribal and religious tags to attract voter support as their political organizations otherwise held little attraction. The candidates were chosen for these personal qualifications (besides their friendships with Babangida): The SDP's candidate Chief Moshood Abiola is the unusual combination of a Yoruba and a Moslem. He is a well-known and wealthy industrialist who may have cut some corners in building his wealth. More importantly, despite his amity with Babangida, he is much less popular with many other senior military men. His NRC opponent, Bashir Othman Tofa, a northerner and a Moslem, was less well known. This face-off assured Abiola of the Yoruba vote in the southwest without ceding support from the Islamic north.[5]

The days leading to the 12 June 1993 contest saw a number of moves and countermoves supporting or opposing the election. They took place mainly in the form of legal actions in state courts (whose decisions were all the more confusing given their multiplicity and co-equal status). On 8 June 1993, a little-known group called the Association for a Better Nigeria won an interim injunction from the Abuja High Court ordering the electoral commission to withhold the results of the upcoming poll. While the Association based its arguments on the prospect for massive voter fraud, its true motivation derived from the wishes of the military, who it claimed were its allies (and their presumed role as a proxy for Babangida). On the eve of the election, the Association won a second order from the Abuja High Court ordering the electoral commission to cancel the vote itself. However, the election did go forward, reputedly following word from the U.S. Government that it would look very unkindly on a cancellation.[6]

As partial results from the 12 June 1993 poll began to filter out, Abiola appeared to have a sizeable lead over rival Tofa. Turnout was low, about 30 percent overall, and perhaps as low as 15 percent in some districts. The larger stature and ethnic-religious attractions of Abiola apparently carried the day. Although charges of vote fraud by the losing party were made, they were not unanimous; more disinterested, international observers did not report "any widespread vote rigging." In fact, *The Economist* editorialized that the election was "reckoned by observers to have been the fairest in most Nigerians' memory." Nonetheless, the National Electoral Commission agreed not to release the vote tallies, thus bowing to the Abuja court order and perhaps also to surrounding troops ostensibly dispatched to protect the Commission.[7]

A Voided Vote

On 23 June 1993, with its results still sealed, the election was annulled by the Babangida government, apparently setting back the timetable for the return of civilian rule for the fourth time in three years.

Shortly thereafter, Babangida first promised that a democratically elected president would be sworn in on the original 27 August 1993 timetable, and then that new elections with different presidential candidates would be held. Babangida based his announcements on the number of conflicting court rulings and on the government's charge that Abiola and Tofa spent $60 million in a spree of vote buying. (A similar corruption charge underlay the annulment of 1992 presidential primaries when some $650 million to $815 million was spent legally and illegally.)[8]

A wide spectrum of opinion decried the failure to publish the election results: student groups, the local Roman Catholic Archbishop, and the Campaign for Democracy, a prominent coalition of forty-two human rights and social organizations (accused by the Babangida government of being a U.S. front).[9] The protests also had a violent aspect that was to foreshadow events over the following months.

In early July 1993, protests again turned violent, particularly in southwestern Nigeria as tens of thousands of Nigerians rioted. In two days, over a hundred people were killed in the violence according to reports by human rights groups. In the largest city of Lagos, rioters barricaded roads, set fires, looted, and beat and stoned police. Normal commerce and transport were shut down until army troops were brought in to assist police in restoring order. The rioting prompted thousands to withdraw their savings and leave the cities for their home towns to avoid the mayhem.[10]

In mid-July, 1993, in an effort to help quell the mounting unrest in the southwest and in recognition of the military's dominant position, the NRC and SDP agreed to a deal with the Babangida government. Given a choice of elections by month's end or the formation of a civilian coalition government, they opted to form an interim national government (perhaps distrustful of the military's appetite for another vote) to be composed of civilian appointees to the presidency and the cabinet. While this effectively voided Abiola's victory, the parties were able to retain the National Assembly and other elected bodies, rather than concede to Babangida's original stipulation that all elected political bodies from the local to national level be dissolved. However, the acquiescence of the parties did not signal acceptance by Abiola; the apparent victor of the June poll sharply criticized the deal and was to later travel abroad to garner support.[11]

Apparently, the agreement did not signal total acquiescence by Babangida either, who was to change his mind twice more. Shortly after the would-be accord with the two parties, he rejected the coalition government as unworkable and called for fresh elections on 14 August 1993. This timetable allowed the parties only about one month to find new candidates (as Abiola and Tofa were banned) and put together their campaigns to install a new president on 27 August 1993. In any event, only the losing NRC agreed with the plan, as the SDP planned to boycott the poll in support of Abiola. Then, in another about-face some two weeks later, Babangida announced renewed support for the agreement to set up a coalition government. He may have been swayed by the parties' pleas that the time was too short to hold elections, or by the threat of renewed rioting and street violence during the polling.[12]

On 26 August 1993, General Babangida kept half his promise to relinquish power to an elected, civilian president. But the "new president" was not the apparent victor of the June, 1993 polling, but Ernest Shonekan, the appointed head of Babangida's Transitional Council and a top executive of United African Co., Nigeria's largest conglomerate. His tenure is supposed to be an "interim" one until new elections can be held in 1994. The Council was set up by Babangida in December, 1992 as a step toward civilian control and earned high praise in financial circles for its abilities in financial matters and good intentions for economic reform. However, even before being tapped as the "interim government," the Council was seen as beholden to the military men who held the true levers of power. While Babangida and most military men resigned from the government, three key cabinet ministries in the Shonekan government were retained by the army or key Babangida allies: defence, information, and capital region (Abuja) administration.[13]

The latest promises did not change the growing conviction that Babangida was using every trick possible to delay a return to civilian rule. His resistance came despite his control of the national electoral commission, the timing of the polls, the creation of the parties, the party platforms, and the pool of potential candidates. In a telling soccer analogy, Richard Joseph, a fellow for African governance at the Carter Center of Emory University and a former lecturer at Nigeria's University of Ibadan, wrote that Babangida "proved a master in drawing civilian politicians into one complicated transitional exercise after another, only to shift the goal posts whenever the politicians seem ready to score."[14]

Perhaps a more salient long-term point made by Joseph is that the military has so insinuated itself into Nigerian politics since 1983, that it will remain the controlling agent regardless of if, when, and how nominal political power is transferred to civilian control.

> Nigerian politics...has become increasingly "Latin Americanized." Even when the military eventually relinquishes its hold on formal political power, it has so entrenched itself as a governing force that it will retain considerable capacity to destabilize and, of course, remove any civilian successor.[15]

Aftermath

One result of the voiding of the June, 1993 results was international censure. Except for humanitarian aid, the United States' government halted all development and security assistance, took back $11 million in assistance to the Ministry of Health, and promised a review of all defense-related exports. Despite its close political and historical ties, the United Kingdom similarly cut off its nonhumanitarian foreign aid. Both governments supported Nigerian protests against Babangida's actions without, noted a *Washington Post* lead editorial, the usual "diplomatic hand-wringing." Despite this firm stand, the financial aid weapon is a less useful short-term lever against Nigeria than it might be in the case of Kenya or Malawi as the $70 million in American and British aid is dwarfed by Nigeria's oil revenues; however, western support for Nigeria's efforts to gain external debt relief could provide sufficient leverage in the medium term.[16]

In an August, 1993 press conference in Washington, D.C., Abiola also called for other nations to withdraw their recognition of the Shonekan government. As strikes continued in his home country, the would-be president also asked the United Nations Security Council to vacate the Nigerian seat and to impose Haiti-style economic sanctions. While Abiola's requests were not honored, he did benefit from a barrage of western media support. Representative headlines included: "Broken Word in Nigeria" (*The Washington Post* editorial, 2 July 1993); "Nonsense in Nigeria" (*The Economist* editorial, 3 July 1993); "Africa's Newest Despot" (*The Washington Post* news analysis, 8 July 1993); and "Babangida Must Go" (*The Economist* editorial, 7 August 1993).[17]

Abiola gathered support on the domestic front as well. He received pledges of support from a diverse and symbolically powerful range of Nigerians. The Sultan of Sokoto, who is the spiritual leader for Nigeria's 40 million Moslems, broke his long-standing support for Babangida by calling for Abiola's installation. Babangida's two military predecessors also came out in favor of Abiola: Buhari and Olusegun Obasanjo (who in 1979 became the only Nigerian military ruler to voluntarily cede power to civilian rule). Perhaps more worrisome to Babangida than the words of retired soldiers was the resignation of Colonel Abubakar Umar and the support of junior officers for defence minister General Sani Abacha, both of whom turned from supporters to public denouncers of Babangida.[18]

Not all of Abiola's support was in the form of words. Nigerian backers have shown a willingness to strike and demonstrate, raising the potential for destabilizing violence. However, a three-day national strike, called in mid-August, 1993 to protest the military regime, displayed the regional appeal of Abiola: the labor action was effective only in the southwest of Nigeria, again shutting down the city of Lagos.[19]

Another three-day strike called in late August, 1993 to protest the installation of the Shonekan government quickly crippled the economy, particularly in the southwest, by shutting down the transportation links. Within three days of a shutdown by workers in the petroleum industry, gasoline stations began to run dry—worsening the usual fuel shortages. Military air traffic controllers and ground crews replaced their striking civilian counterparts at the country's seventeen airports, but few airlines were able to operate. The principal ports of Lagos and Port Harcourt were hit by a dock workers' strike.[20]

Damaged Reputation

Once the dust settles around the transfer of power, the most lasting damage may be to Nigeria's reputation. Rather than view Nigeria as the vanguard of an emerging Africa, observers may begin to see more parallels with Zaire, a large, populous, and resource-rich African nation which, in the words of *The Economist* in discussing Nigeria, has become "a bad joke." The editors of that magazine continued to describe the Nigerian elections:

> To call an election, allow one's friends to try to get it stopped in the courts, let it proceed nonetheless, annul the result, then announce a fresh version with fancy new conditions to rule out candidates one dislikes—all this sounds more like a comic novel...than serious modern government.[21]

The repudiation of the democratic process was especially telling because of Nigeria's prominent place on the continental stage and

Babangida's repeated public support for the move away from dictatorship in Africa. An African expert at the Aspen Institute in Washington, D.C., Pauline Baker explained,

> Nigeria has gotten so much publicity and represents such a bell-wether in Africa, and has even touted itself as an example of democratization.... Having advertized itself so much as going back to civilian rule, and now having things go so badly, it certainly does plunge people's hopes all over Africa.... It's really a tragedy for all Africa.[22]

Economic Reform

Economic Decline

Nigeria's economic mismanagement led to the embarrassing drift from the ranks of the World Bank's middle-income to low-income countries. As late as 1986, Nigeria showed a per capita GNP of $640, placing it between "MICs" Zimbabwe and Dominican Republic by that measure. Since that time, Nigeria has slipped in the ranking to a 1991 per capita output of $340.[23] This is below LICs Haiti and China at $370, despite the turmoil and poverty that has kept Haiti the poorest nation in the Americas. Perhaps more pertinent is that this "regional superpower" ranks below several nations in West Africa: LICs, the Gambia ($360), Benin ($380), Ghana ($400), Togo ($410), Guinea ($460), and MICs, Cote d'Ivoire ($690) and Senegal ($720). In addition, its African colleague in OPEC, Gabon, ranks in the upper tier of MICs with a $3,780 per capita income on an admittedly more modest population base (1.2 million versus Nigeria's 99 million).[24]

In 1986, Nigeria implemented an International Monetary Fund-style (IMF) Structural Adjustment Program (SAP). Largely for political reasons, the government did not initiate the SAP under IMF auspices, but its stated goals included the typical prescriptions contained in IMF stabilization programs: removal of most price controls, liberalization of foreign exchange and international trade rules, depreciation of the currency (the *Naira*), reform of the financial sector, and the reform or sale of public-sector companies. Later, it agreed to a formal IMF stabilization program but continued its practice of not drawing (borrowing) on IMF stand-by funds. Despite the generally admirable goals of these programs, the results have been mixed at best, and Nigeria is not in compliance with its IMF targets.[25]

On the eve of the August, 1993 "transition" to the Shonekan government, *The Economist* summed up the state of the Nigerian economy this way:

> [A] budget deficit amounting to 10% of GDP; an external debt the servicing of which consumes a quarter of export earnings; inflation estimated in June [1993] to have reached an annual rate of 70–100%; and real income per head a tenth of what it was when [Babangida] took office.[26]

The main underlying causes of this decline are an over-reliance on petroleum as the economic engine, the lack of will to carry through with economic goals and policies, limited progress on economic reform (subsidy reduction, efficient use of private Nigerian and foreign investment, ending wasteful spending on glamorous mega-projects), corruption, and the resulting foreign debt burden.

Failure of the economic stabilization program will discredit reform and open the door for policies that will be even more harmful to Nigeria. For example, the accepted economic wisdom of popular politicians and pundits calls for a managed (rather than market-driven) exchange rate, a distaste for foreign capital and investment, a stance against taxes or borrowing revenues (which "certainly" would be squandered by the government), and a continuation of the belief that Nigerians deserve the cheapest fuel in the world as their share of the nation's petroleum riches.[27]

Over-Reliance on Petroleum

Nigerian dependence on the petroleum sector made it vulnerable to swings in that volatile commodity market. With the first OPEC price shock, Nigeria's share of government revenue derived from oil rose to 82 percent from 26 percent. Despite the mid-1980s drop in oil prices, petroleum receipts are still estimated to account for 97 percent of export income and around 80 percent of federal government revenue in 1991.[28]

The reliance on petroleum, and the belief that it is a national resource to be shared on an individual rather than a national level, hurts the government's efforts to hold the line on spending. Nigeria is hard pressed to loosen its spending policies every time the oil market firms; there is a strong public sentiment to treat each price rise as a permanent change in the market, even when the improvement appears to be truly transitory, such as the run up following the Gulf War. As a result, spending is ratcheted upward and any windfall oil receipts cannot be used to meet stabilization goals (e.g., to smooth out the effect of downward price movements). The 1990 Gulf War-driven spending increases helped push the 1991 budget deficit to a then-record N$35.5 billion (12.4 percent of GDP). As a Morgan Grenfell publication pointed out, "In bald fact, Nigerian fiscal policy has tended to amplify, rather than contain, the unavoidable distortions caused by oil price volatility."[29]

Inability to Carry Through with Financial Goals

Unlike Mexico, which toughed out a weakened international oil market, economic stabilization program, and debt rescheduling, Nigeria was unable to hold to its macroeconomic goals. As *The Economist* opines, "Nigeria's problem is not a lack of prescription, but of political will."[30]

In 1993, the budget put forward by Babangida through the Transitional Council renewed solemn pledges to conform to the stabilization program. However, the announcement of the 1992 results shortly thereafter cast further doubt on the government's allegiance to its own policies: the 1992 budget deficit sunk to a new record of N$44.2 billion versus a projected surplus of N$2 billion. Inflation (consumer prices) was 45 percent propelled by a 41 percent rise in money supply.[31]

The inflation weakened efforts to support the value of the Naira following its devaluation in late 1992; between defending the value of the currency, financing the peace-keeping effort in Liberia, buying back some $3.4 billion of rescheduled foreign debt at a discount, and suffering continued capital flight, Nigeria's foreign exchange reserves plunged from $4.4 billion to $967 million. As the London *Financial Times* noted, in late 1992, the Naira devaluation "had been accompanied by pledges of fiscal restraint, prudent monetary policies, public sector rationalization, and curbs on state spending. The 1992 [budget] deficit speaks for itself."[32]

Limits to Liberalization

Despite some positive steps, the process of liberalization has been slow and incomplete.

Subsidies. One area is subsidies, which have tended to benefit the politically powerful urban third of the population at the expense of rural Nigerians and the petroleum sector. The agricultural population saw its income restricted by price controls on foodstuffs and by an exchange rate that earned it fewer Nairas than would a market rate of exchange. These disincentives contributed to the decline in the agricultural sector in general and cash crops of cocoa, rubber, oil palm, groundnuts, and cotton in particular.

The petroleum industry continues to struggle to meet domestic demand for gasoline at an official price of at N$0.7 ($0.03) per liter. The oil companies, which rarely produce enough additional petroleum to fill the nation's OPEC export quotas, require about $500 million to $1 billion in annual subsidies from the government to remain viable as local prices are below production costs. Overall the subsidy costs the government about N$40 billion even after factoring its own spending on fuel; this amount is almost enough to balance the N$44 billion federal budget deficit all by itself. Besides forgoing huge export and government revenue, the government's sale of cheap oil domestically encourages waste at the expense of potential exports, the importation of automobiles, and the smuggling of the cheap oil into neighboring countries (such as Benin where the retail price is N$9.7/liter) for sale at world prices, which has even led to internal shortages.[33]

There are dire predictions that unless the government can keep oil prices at realistic levels, the subsidized consumption and smuggling of petroleum may deplete this resource within a generation. The Babangida government has done away with many price subsidies and has agreed in principal with the imperative of raising petroleum prices; in practice, however, it has been very careful in toying with the gasoline subsidy. One reason is its memory that a 3 percent April, 1988 price increase had to be reversed after it sparked nationwide riots. With popular fears that any increase would be siphoned off for corrupt officials, trade union and student leaders have pledged public actions if prices are raised.[34]

Privatization and foreign investment. Another area of liberalization that has shown mixed results is privatization. This effort has yet to make a dent in the largest and most important state firms, such as Nigerian Airways, Nigerian National Shipping, National Paper Manufacturing Company, six steel mills, and six vehicle assembly plants.[35]

Defenders of the program point out that in its five-year tenure the Technical Committee on Privatization and Commercialization (TCPC) properly disposed of the 111 enterprises identified. TCPC effectively worked itself out of existence and was replaced in 1993 with the Bureau for Public Enterprise, whose goal is to monitor performance of the remaining parastatals. The fifty-five companies actually sold generated N$3.3 billion in sales receipts, about five times the original government investment, and created as many as 800,000 new shareholders. Of the remaining fifty-six firms targeted, eighteen were privatized before TCPC was set up, nine were to be privatized shortly after TCPC's dissolution, four have been commercialized, another eight cannot be sold as they are technically insolvent, the six money-losing vehicle assembly plants were not deemed salable, and eleven firms did not merit further action.[36]

However, critics point out that the most important firms have yet to be touched, parastatals are still losing money, and the total government investment in the sector continues to grow despite the sell-offs.[37] Commenting on the TCPC report that work was completed on eighty-eight of the 111 firms, *The Economist* argued:

> This sounds respectable, except that the 111 left out all the important ones: Nitel (the telecommunications monopoly), NEPA (the electricity monopoly), the Nigerian Railway Corporation, NNPC, Nigeria Airways, Nigeria Ports, cement works, fertiliser plants, flour mills. The same report went on to say that the total value of the enterprises sold so far amounted to less than 2% of the government's portfolio.[38]

Another aspect of liberalization is the willingness and ability to attract foreign capital. The petroleum sector needs sizeable investment, especially to tap Nigeria's long-neglected natural gas reserves. A liquified natural gas (LNG) project has been in the talking stage for three decades, but tangible progress is yet to come in creating this facility.

Wasteful glamour projects. With the oil boom, Nigeria was encouraged to think of itself in grand terms. While many of the projects being privatized may have seemed good ideas at their inception, the underlying assumptions were proved incorrect or the subsequent performance was disappointing. However, other projects were obviously wasteful or economically unfeasible even at the outset.

These white elephants accounted for large portions of debt for projects of questionable economic value. They are epitomized by the government's attempts to create a national steel industry and construct a new national capital city. The Ajaokuta steel mill is a twenty-year project designed to produce steel in the Nigerian interior. However, with no high-grade coal or iron ore in Nigeria, the mill faces huge transportation costs, which would make it cheaper to simply import the metal.[39]

Abuja is a Brasilia-like project constructed in an ethnically "neutral" site in central Nigeria. The construction of Abuja was decided in 1975 to escape the unlivable aspects of Lagos and to build a truly national capital city. Progress has been fitful, dependent on oil revenues, but the city now houses many modern buildings including a presidential palace, an international conference center, and headquarters for the SDP and NRC. While congressmen elected in 1992 have stayed at a local luxury hotel, few government agencies have relocated because of a lack of housing. As a result, the new capital is largely deserted.[40]

Other projects included a Nigeria-to-Benin divided expressway, which is now filled with potholes, and the Festac model housing project in Lagos, which suffered from a typhoid outbreak when sewer pipes burst and contaminated drinking water.

Corruption

The use of political office for personal gain is endemic in Nigeria. This is shown not only in the prevalence of corruption on an individual level, but also in the promulgation of economic policies that permit the enrichment of an elite at the expense of the entire nation. While corruption in itself may not be unusual, the existence of the petroleum boom added two singular factors: The revenues generated by Nigeria's petroleum exports made the potential scope of the corruption enormous. The "unearned" nature of the petroleum wealth provided a convenient excuse for a profligate use of the proceeds.[41]

Some of the opportunities for wide-scale theft by dishonest officials and businessmen have been eliminated along with the numerous trade restrictions and the exchange rate controls, which they could manipulate to their personal advantage. But opportunities for graft still exist, such as the gasoline subsidy's encouragement of smuggling to neighboring countries, or white elephants.

Another factor making theft feasible is that many financial data are not available for analysis. This includes the costs of the construction of Abuja, the tab for peacekeeping in Liberia, or, most importantly, the accounts of the Nigerian National Petroleum Corporation (NNPC). Without these figures, analysts are left with only indirect measures—which lead to estimates that as much as $2.7 billion (10 percent of GDP) is being siphoned off from NNPC annually.[42]

The costs of this corruption are a reduction in economic efficiency, the diversion of government revenues, and the damage to political credibility. This last point is amply illustrated by the quandary over the gasoline subsidy: If Nigerians believed that the revenues from the higher gasoline price would accrue to the public good rather than the politicians' pockets, the gasoline subsidy could have been cancelled years ago.[43]

External Debt Pressures

While the slashing of Nigerian oil receipts that accompanied the post-OPEC II oil glut was an important cause of its debt servicing difficulties, the problem was compounded when the capital was directed to economically wasteful purposes that did not generate any enhanced earnings capacity.

For Nigeria, the package of domestic policies was a key contributor to its debt difficulties. Subsidized prices and an overvalued exchange rate encouraged consumption of imports while discouraging production for export. What foreign exchange did accrue from exports and borrowings was used to support subsidies, corruption, and wasteful public projects rather than address the underlying problems. In a region characterized by nations without the means to improve their lot, Nigeria had the means, but chose to use them unwisely.

The ensuing debt burden has put a strain on the government's room for maneuver. Interest on these obligations was around 7 percent of GDP in 1986–91, roughly equal to the federal capital investment budget. With 99 percent of the $34 billion in reported foreign debt owed or guaranteed by the public sector, foreign debt service payments also represented about 60 percent of federally retained revenues (after the automatic distribution of about 44 percent of receipts to the states). Rescheduling of the debt owed to foreign commercial banks along the lines of the so-called "Brady Plan" allowed for a reduction of principal or interest and a lengthening of maturities, but the $5.6 billion affected by this "London Club" accord was only about one-sixth of the total debt burden.[44]

The bulk of the debt is extended by official creditors. These government and multinational lenders would probably be amenable to a significant easing of debt and debt service burdens on humanitarian and practical grounds. However, the "Paris Club" of western governments first would require that Nigeria negotiate and comply with stabilization programs with the IMF. This would ensure that the debt relief contributes to a permanent solution to Nigeria's debt crisis. While the largest creditor, the United Kingdom, may be willing to back concessional relief for Nigeria on faith, London would need a "tough and comprehensive" IMF adjustment program and a track record of compliance to convince less forgiving creditors, including Germany and Japan.[45]

A Paris Club agreement is critical because had Nigeria paid all the debt service due to all foreign creditors, those payments would have taken up 133 percent of federal revenues. The potential gains for Nigeria are all the greater with its official classification as a heavily indebted, low-income country, which makes it eligible in theory for concessionary rescheduling terms. According to provisions agreed on by the Paris Club in December, 1991 in Toronto, these would include reductions in debt service, cancellation of debt, and extensions of repayment periods. Per World Bank estimates, the combination of these concessions would be sufficient to reduce Nigeria's foreign debt to $16 billion from $29 billion, and its debt service ratio to 21 percent in 1993–96 from 36 percent.[46]

Implications of Failed Policies

The economic difficulties have forced Nigeria to make painful sacrifices elsewhere. On the geopolitical front, on 31 August 1993, Shonekan decided to withdraw Nigerian troops from the West African peace-keeping force in Liberia by March, 1994. Nigeria was the largest of the six coalition members, providing 9,000 of the 14,000 soldiers sent to the war-racked nation in August, 1990 under the auspices of the Economic Community of West African States (ECOWAS). Shonekan cited the financial costs of the commitment. This step was critical for both Liberia and Nigeria. For Liberia, the war—which has cost 150,000 lives—was suspended by a cease-fire agreed upon in July, 1993. The truce's fragility was made manifest by the facts that disarmament of the Liberian factions had not begun by September, 1993, and that rebel leader Charles Taylor boycotted the first meeting of the transitional government. All humanitarian concerns aside, for Nigeria, its withdrawal announcement at this critical juncture punctured its claims of local superpower status and its continued domination of ECOWAS.[47]

Agriculture, Energy, and Natural Resources

Nigeria's nonpetroleum resources have fallen victim to the country's focus on its oil riches.

Agriculture

The agricultural sector has also suffered in the last three decades resulting in Nigeria's conversion from a significant net food exporter to importer. A key problem was the long-time overvaluation of the Naira, which made Nigerian produce uncompetitive at home and abroad compared with foreign foodstuffs. The economic reforms of the late 1980s appear to have stopped—but not reversed—the trend. Until the Naira devaluation in 1992, the government's "solution" was to further distort the market by banning trade in certain goods, such as a prohibition on wheat imports.[48]

Nigerian agriculture is dominated by small farmers who account for 90 percent of production. This form of agriculture has been favored by World Bank and other international development programs, but is not without problems such as deforestation, erosion, and soil depletion. Simplification of some government policies under the reform program has helped, such as the removal of government corporations from the agricultural marketing. However, there are such a large number of government initiatives and policies that at times they inevitably work at cross purposes. For example, the government is also promoting large-scale agriculture through the importation of machinery and the construction of dams.[49]

An example of both the positive effects of liberalization and the pitfalls of excessive governmental interference is cocoa production. Mirroring in large part the entire agricultural sector, cocoa declined as a major crop in Nigeria. However, reforms begun in 1986 have helped boost production and exports, despite lower international prices. In particular the dissolution of the agricultural marketing board allows producers to obtain the highest price from exporters or licensed buying agents (and the Naira devaluations have offset lower world prices). The 1992–93 cocoa crop estimate of 135,000 tonnes is double the output of pre-reform 1986, and stands to generate $90 million in foreign exchange earnings, making it the largest export except for petroleum.[50]

Despite this success in nurturing a traditional agricultural sector and diversifying Nigeria's economic and export base, the government is considering rules to require the local processing of the cocoa. By banning or taxing the export of raw beans, Nigeria hopes to further develop and diversify its economy by encouraging the formation of a local processing industry. However, even from a national viewpoint, this may not be the best move: gains from the extra value-added output could be offset by the costs of machinery and by employment losses as the industry mechanizes. Internationally, the weak cocoa market has hurt capacity and profitability at even the most efficient European plants. The idea of the raw cocoa ban has been bandied about for several years, and its continued life has discouraged farmers from making the long-term investments in fertilizer use and hybridization that could lead to even further boosts in cocoa output.[51]

Energy

An example of the complacency engendered by the discovery of oil is the two-decade-old failure to develop Nigeria's natural gas resources. It is agreed that its vast natural gas reserves would generate additional foreign exchange, attract foreign investment, and diversify export earnings. The typically longer-term pricing structure of LNG contracts (which require significant infrastructural investments) would be a welcome offset to the often erratic behavior of the international crude oil market.

Nigeria's proven reserves are 2.6 trillion cubic feet (tcf) and additional probable reserves are 1.8 tcf. The proven resources are equivalent to eleven times the amount of natural gas used annually in the United Kingdom, Germany, France, Italy, Belgium, and Spain.[52]

There is a long-held consensus that export of the resource in the liquified LNG form is more workable than the construction of a pipeline to prospective customers in Western Europe. The export of LNG requires significant investment: pipelines to transport the gas internally, a liquefaction facility in Nigeria, a terminal for transferring the LNG to special ships, the LNG ships themselves, and unloading and processing terminals in the purchasing countries.

After some twenty years of indecision, it appeared that an LNG project would be approved in late 1992: Nigeria LNG bought four ships, signed purchase contracts, and began preliminary work on one of two proposed

LNG terminals. However, the Nigerian government showed its displeasure with the contract awarded by technical adviser, Shell Gas, in October, 1992 by dissolving the board of NLNG, allowing contractual deadlines to lapse, and thereby adding another chapter to Nigeria's failure to take advantage of this resource. In contrast, fellow oil exporter, Indonesia, has constructed five LNG projects in this same time span.[53]

Nigeria's proven oil reserves can cover current production levels for twenty-seven years. These reserves should be augmented by renewed exploration as oil companies enter (Exxon), return (British Petroleum, for the first time since 1979), or continue (Shell Oil) their activities. They have been encouraged by government reforms that no longer require them to undertake joint exploration ventures with the chronically un- or underfunded NNPC.[54] Unfortunately, Nigeria's upstream operations are not looking as positive. Its refining industry, like many parastatals, is highly inefficient mainly because of the absence of a penalty for losing money. Foreign oil companies judge that the four Nigerian refineries have five times as many workers as necessary. In 1992, one refinery was shut down due to a fire, two others had their output drop by more than 20 percent, and only the fourth registered an improvement.[55]

Political Centralization

The borders of most African states reflect colonial frontiers rather than historic tribal or ethnic groupings. Nigeria proved that it is no exception when it underwent the costly Biafran civil war. That war was preceded by the 1964 election, which was boycotted in the east and west of Nigeria; with such a tarnished mandate, that government lasted only eighteen months. Another catalyst was the 1966 massacres of Ibos (the eastern tribe) in northern Nigeria. In 1967 the east, naming itself the new nation of Biafra, seceded and sparked a three-year civil war. After suffering about 1 million deaths both from the fighting and famine, the nation was forcibly reunited.

Nigeria has three dominant groups with tribal, geographic, and religious identities, although throughout the country there are hundreds of smaller tribes who have formed coalitions to rule several of Nigeria's twenty-one state governments. Under British colonial rule, the principal divisions were accented as London split the colony into three regions and fostered (or created) local chiefs. However, with the federal victory in the Biafran war, the nation remained united in contrast to other British federations such as Southern Rhodesia, Northern Rhodesia, and Nyasaland, which split to become the independent states of Zimbabwe, Zambia, and Malawi.[56]

The north of the country is overwhelmingly Moslem and home to the Hausa-Fulani group, whose members spill over into neighboring Niger and Benin. The Hausa-Fulanis have dominated the military and, not coincidentally, have governed Nigeria since its independence in 1960. Given the prevalence of political favoritism and corruption, this means that northerners have also reaped most of the tangible benefits of governmental largesse.[57]

The Yoruba is the primary group in the southwest—including the largest city of Lagos—and would-be president Abiola is a member. Its leaders are primarily Christian, although Abiola's Moslem beliefs helped him garner voter support in an unprecedented fashion across religious, tribal, and regional lines. His victory began to blur the regional identities of the parties—SDP for the south and NRC for the north.

The southeast, the would-be Biafra, is home to the Ibos. Its inhabitants are predominantly Animist and Christian. Their political traditions have been looser than the northerners, and they more readily adapted to western customs and religion. Besides the potential for bitterness from the Biafran conflict, Moslem radicalism, and northern political dominance, this region is home to Nigeria's oil reserves: This has given rise to the claim that the southeast produces Nigeria's wealth, and the northern politicians spend it.

The Babangida regime's refusal to accept Abiola's victory in the June, 1993 presidential poll has allowed the debate over democratic transition to revert into a largely tribal or ethnic conflict. The head of a local civil rights organization opined:

> The message is that no southerner can be president of Nigeria. If this continues, the divisions will become north versus south. The big northern aristocracy that has dominated the political space simply is not ready to see a southerner elected because of the north's reliance on the spoils of political patronage.[58]

Violence in Nigeria seems to be on the rise: between the economic, tribal, religious, and, as discussed above, political groups. Two incidents occurred in April, 1991 in northern Nigeria. In the town of Katsina,

Shiite Moslem militants began to set fire to government buildings in an organized effort before being faced by government troops. The action was in response to the military governor's threats against a radical clergyman who was the local representative of Ibrahim el Zakzaky, dubbed by newspapers the "Ayatollah of Nigeria."[59]

Later that month, in Bauchi, a bloodier fight occurred between Christians and Moslems over the use of a local slaughterhouse that would have rendered it unclean under Islamic law. In the first wave of protests, dozens were killed. In response, the Moslems took over the town for two days, killing hundreds of Christians, inspiring thousands more to flight, and torching offending buildings. When the army moved in to reclaim the town, at least 120 more Moslems were killed before the military restored order. The death toll in Bauchi may have easily exceeded 1,000.[60]

In October, 1991 more Moslem-Christian violence flared up in the northern city of Kano, less than a week before primary elections. In a Christian district, ostentatious preparations for a religious rally prompted a Moslem attack. The fighting lasted two days and cost hundreds of lives, with both sides relatively even in strength. In the wake of the fighting, many Ibos fled south, prompting memories of the Ibo massacres that preceded the Biafran conflict.[61]

Later that month, troops were called into the provincial town of Zangon Kadif after at least several hundred Hausa and Kataf people died in a dispute over the location of a marketplace. What began as more of a localized ethnic conflict between the groups was quickly transformed into a religious battle between Christians (Katafs) and Moslems (Hausas) in Zangon Kadif and neighboring towns.[62]

Nigeria's Islamic population, at least 43 million strong, seems to have grown less tolerant and more militant in recent years. Influence over the community is shifting from the moderate Sultan of Sokoto to Sheik Zakzaky, who advocates an Iran-inspired Islamic republic of Nigeria. In this more radicalized atmosphere, more than 3,000 people were killed in fights with Christians or other sects in 1988–93, and the practice of burning down "offensive" Christian churches has become more and more common in the north.[63]

More recently, violence has political dissatisfaction at its root, but Nigeria has been racked by riots and strikes protesting a variety of perceived injustices. In May, 1992 Lagos saw violent outbreaks in reaction to fuel shortages and increasing poverty. Rising violence and tension were behind the government's decree that same month banning any religious, political, or ethnic association outside of the two official political parties upon pain of a three-year incarceration.

The government has tried to foster a true Nigerian identity: making English the official language, ensuring that all military units were geographically mixed, and requiring that all young people serve elsewhere in the country on national youth service.[64] However, without an ability to convince the majority of the people that policy is pursued in the best interests of the nation and all its peoples, the government's efforts to reform the economy will be hamstrung by public unrest each time an unpleasant choice needs to be made. In the long run, the simple existence of the nation may rely on all Nigerians believing that they have a stake and a voice in Nigeria. Without this national will, Nigeria could degenerate into the civil war that has virtually destroyed the nearby countries of Chad, Sierra Leone, and Liberia.

Conclusion

Nigeria is sub-Saharan Africa's best hope for a tiger to emerge in the coming decade. It has the resources—human and natural—to effect the transformation, but it needs to act decisively, and soon, to reverse the disastrous slide into the ranks of poor nations that began in the 1980s. It can emulate Mexico and move ahead, or Zaire and slip into obscurity.

The key will be the installation and survival of a democratic government and the development of a true democratic tradition. That government must be popularly based, because it is critical that it be perceived as answerable and loyal to the Nigerian people. Without that perception, it will be unable to effect the tough policy choices required, such as ending the gasoline subsidy, privatizing the major parastatal money losers, bringing government spending into line, and keeping the foreign exchange rate competitive.

Perhaps as important as all of these specific policies is the need to foster the public belief that corruption will not be tolerated, and that any sacrifices asked of the Nigerian people will truly make the nation stronger, not just some corrupt politicians richer. Unless it is "squeaky clean," the government will find it impossible to overcome the cynicism that has taken root in an Nigerian electorate that has watched its petroleum patrimony plundered.

If this democratic government can turn the Nigerian economy around, it will be in a position to build on this success to argue its case that Nigeria is greater than the sum of its parts. This is the only sustainable basis to overcome the ethnic, geographic, and, above all, religious divisions that have flourished in the past several years. Nigeria will either be a democratic tiger, or it will remain an autocratic elephant.

Notes

1. "Anybody Seen a Giant?," *The Economist* (21 August 1993): S4.
2. This section adapted in large part from David L. Crum, "Sub-Saharan Africa," *The Global Debt Crisis: Forecasting for the Future,* eds. Scott B. MacDonald, Margie Lindsay, and David L. Crum (London: Pinter Publishers Limited, 1990).
3. "Voting Against the Odds," *The Economist* (19 June 1993): 42.
4. Ibid., 42.
5. Ibid., 42. Cindy Shiner, "Nigerian Cancels Vote, Agrees to Interim Rule," *The Washington Post* (1 August 1993): A24. Paul Adams, "Nigeria's Other Head of State," *Financial Times (London)* (26 July 1993): 6.
6. "Voting Against the Odds," 42. On 17 July 1993, Abimola Davis, a top executive for the Association for a Better Nigeria, admitted that the suits were filed to buy Babangida several more years in power. Reporters were detained for forty minutes after the press conference to allow Davis and his family to leave Nigeria. Cindy Shiner, "Nigerian: Suits Filed to Stall Vote," *The Washington Post* (17 July 1993): A10.
7. Michelle Faul, "Nigerian Groups Demand Election Results Be Released," *The Washington Post* (18 June 1993): A34. "Voting Against the Odds," 42. "That's Democracy," *The Economist* (26 June 1993): 43. "Babangida Must Go" (Editorial), *The Economist* (7 August 1993): 18.
8. Michelle Faul, "Nigerians Assured of a President in August," *The Washington Post* (26 June 1993): A20. Associated Press, "Nigerian Leader Promises One More Try at Elections," *The Washington Post* (27 June 1993): A28. "General Babangida's Twisting Road to Democracy," *The Economist* (24 October 1992): 43.
9. Faul, "Nigerian Groups Demand Election Results Be Released," A34. Faul, "Nigerians Assured of a President in August," A20.
10. Mark Fritz, "11 Die as Nigerian Rioters Protest Military Rule," *The Washington Post* (7 July 1993): A23, A28. Cindy Shiner, "Nigeria Muzzles the Press," *The Washington Post* (24 July 1993): A13. "A New Glove," *The Economist* (28 August 1993): 40.
11. Reuter, "Nigerian Parties Agree to Interim Government," *The Washington Post* (8 July 1993): A11. "Nigerian Reject[s] Deal," *Christian Science Monitor* (9 July 1993): 20. "Nigeria's Next Steps" (Editorial), *Christian Science Monitor* (9 July 1993): 18. Associated Press, "Nigeria's Would-Be Ruler Rejects Deal," *The Washington Post* (9 July 1993): A22.
12. "Round in Circles," *The Economist* (17 July 1993): 36–37. Shiner, "Nigerian Cancels Vote, Agrees to Interim Rule," A24.
13. Cindy Shiner, "Nigeria's Military Ruler Resigns, Appoints Interim Government," *The Washington Post* (27 August 1993): A27–28. "Promises, Promises," *Financial Times (London)* (1 April 1993): I.

14. Richard Joseph, "Africa in the Latin Style," *The Washington Post* (6 July 1993): A15.
15. Ibid.
16. Andy Mosher, "Nigerian Urges U.S. Not to Recognize New Regime," *The Washington Post* (31 August 1993): A15. "Broken Word in Nigeria" (Editorial), *The Washington Post* (2 July 1993): A18. Keith B. Richburg, "Africa's Newest Despot," *The Washington Post* (8 July 1993): A11.
17. Mosher, "Nigerian Urges U.S. Not to Recognize New Regime," A15.
18. "Round in Circles," 36–37. "A New Glove," 39.
19. "Strike Protesting Army Rule Shuts Down Nigerian Capital," *The Washington Post* (13 August 1993): A30.
20. Associated Press, "Strikes Ground Airliners, Cause Gasoline Shortages," *The Washington Post* (31 August 1993): A15. "Change, of a Sort," *The Economist* (4 September 1993): 42.
21. "Nonsense in Nigeria" (Editorial), *The Economist* (3 July 1993): 20.
22. Richburg, "Africa's Newest Despot," A11.
23. Press reports show the per capita figure continuing to drop to $250 in 1993, from $1,000 tens years earlier. Shiner, "Nigeria's Military Ruler Resigns," A28.
24. World Bank, *World Development Report 1988* (New York: Oxford University Press, 1988), 222. World Bank, *World Development Report 1993*, 238–39.
25. Morgan Grenfell Debt Trading and Arbitrage, "Nigeria: Prospects for Debt and Adjustment," *Developing Country Research* (22 September 1992): 3.
26. "Anybody Seen a Giant?" S4.
27. Tony Hawkins, "Quick Fix Solutions Not on Offer," *Financial Times (London)* (1 April 1993): III.
28. Morgan Grenfell Debt Trading and Arbitrage, "Nigeria: Prospects for Debt," 5. "Oiling the Wheels," *The Economist* (21 August 1993): S4.
29. Morgan Grenfell Debt Trading and Arbitrage, "Nigeria: Prospects for Debt," 5–6.
30. "The Economic Puzzle," *The Economist* (21 August 1993): S7.
31. "Promises, Promises," I. Inflation, money supply, and reserves data updated by International Monetary Fund, *International Financial Statistics, August 1993* (Washington, D.C.: International Monetary Fund, 1993), 398.
32. "Promises, Promises," I. Inflation, money supply, and reserves data updated by International Monetary Fund, *International Financial Statistics,* 398.
33. Morgan Grenfell Debt Trading and Arbitrage, "Nigeria: Prospects for Debt," 6. Tony Hawkins, "A Tighter Rein on Public Spending," *Financial Times (London)* (1 April 1993): IV.
34. Morgan Grenfell Debt Trading and Arbitrage, "Nigeria: Prospects for Debt," 6. "Promises, Promises," I.
35. Morgan Grenfell Debt Trading and Arbitrage, "Nigeria: Prospects for Debt," 6.
36. Tony Hawkins, "Focus is on Performance," *Financial Times (London)* (1 April 1993): VI.
37. Ibid., VI.
38. "Breaking the Cycle," *The Economist* (21 August 1993): S9.
39. "The Economic Puzzle," S7.
40. "Breaking the Cycle," S11.
41. "Oiling the Wheels," S4–5.
42. "The Economic Puzzle," S8.
43. "Oiling the Wheels," S4–5.
44. Morgan Grenfell Debt Trading and Arbitrage, "Nigeria: Prospects for Debt," 7.

References

Books

International Monetary Fund. *International Financial Statistics, August 1993*. Washington, D.C.: International Monetary Fund, 1993.

MacDonald, Scott B., Margie Lindsay, and David L. Crum, eds. *The Global Debt Crisis: Forecasting for the Future*. London: Pinter Publishers Limited, 1990.

World Bank. *World Development Report 1988*. New York: Oxford University Press, 1988.

______. *World Development Report 1993*. New York: Oxford University Press, 1993.

______. *World Debt Tables, 1992*. Washington, D.C.: World Bank, 1993.

Periodicals

Adams, Paul. "Analysts Urge More Liberalization." *Financial Times [London]* (1 April 1993): IX.

______. "Doubts About Local Processing." *Financial Times [London]* (1 April 1993): IX.

______. "Nigeria's Other Head of State," *Financial Times [London]* (26 July 1993): 6.

Associated Press. "Nigeria's Would-Be Ruler Rejects Deal." *The Washington Post* (9 July 1993): A22.

______. "Nigerian Leader Promises One More Try at Elections." *The Washington Post* (27 June 1993): A28.

______. "Strikes Ground Airliners, Cause Gasoline Shortages." *The Washington Post* (31 August 1993): A15.

"Attention." *The Economist* (26 October 1991): 52.

Balls, Edward. "Debilitating Burden Grows." *Financial Times [London]* (1 April 1993): V.

______. "Tough Talk by Creditors." *Financial Times [London]* (1 April 1993): V.

"Breaking the Cycle." *The Economist* (21 August 1993): S8–10.

"The Call of Islam." *The Economist* (21 August 1993): S12–13.

"Change, of a Sort." *The Economist* (4 September 1993): 42.

Christian Science Monitor (9 July 1993).

"The Economic Puzzle." *The Economist* (21 August 1993): S6–8.

The Economist (3 July 1993).

______. (7 August 1993).

Faul, Michelle. "Nigerian Groups Demand Election Results Be Released." *The Washington Post* (18 June 1993): A34.

______. "Nigerians Assured of a President in August." *The Washington Post* (26 June 1993): A20.

Fritz, Mark. "11 Die as Nigerian Rioters Protest Military Rule." *The Washington Post* (7 July 1993): A23 ff.

"General Babangida's Twisting Road to Democracy." *The Economist* (24 October 1992): 43–44.

"Godly Slaughter." *The Economist* (4 May 1991): 47–48.

Hawkins, Tony. "Focus is on Performance." *Financial Times [London]* (1 April 1993): VI.

______. "Quick Fix Solutions Not on Offer." *Financial Times [London]* (1 April 1993): III.

______. "A Tighter Rein on Public Spending." *Financial Times [London]* (1 April 1993): IV–V.

Joseph, Richard. "Africa in the Latin Style." *The Washington Post* (6 July 1993): A15.

Morgan Grenfell Debt Trading and Arbitrage. "Nigeria: Prospects for Debt and Adjustment." *Developing Country Research* (22 September 1992).

Mosher, Andy. "Nigerian Urges U.S. Not to Recognize New Regime." *The Washington Post* (31 August 1993): A15.

"A New Glove." *The Economist* (28 August 1993): 39–40.

"Nigerian Reject[s] Deal." *Christian Science Monitor* (9 July 1993): 20.

"Oiling the Wheels." *The Economist* (21 August 1993): S4–5.

Pilling, David. "Dream in the Balance." *Financial Times [London]* (1 April 1993): VII.

"Promises, Promises." *Financial Times [London]* (1 April 1993): I.

Reuter. "Nigerian Parties Agree to Interim Government." *The Washington Post* (8 July 1993): A11.

Richburg, Keith B. "Africa's Newest Despot." *The Washington Post* (8 July 1993): A1 ff.

"Round in Circles." *The Economist* (17 July 1993): 36–37.

"Seething." *The Economist* (23 May 1992): 46.

Shiner, Cindy. "Nigeria Muzzles the Press." *The Washington Post* (24 July 1993): A13.

______. "Nigeria Plans to Withdraw from Liberia." *The Washington Post* (1 September 1993): A26.

______. "Nigeria's Military Ruler Resigns, Appoints Interim Government." *The Washington Post* (27 August 1993): A27–28.

______. "Nigerian: Suits Filed to Stall Vote." *The Washington Post* (17 July 1993): A10.

______. "Nigerian Cancels Vote, Agrees to Interim Rule." *The Washington Post* (1 August 1993): A24.

"Strike Protesting Army Rule Shuts Down Nigerian Capital." *The Washington Post* (13 August 1993): A30.

"A Survey of Nigeria." *The Economist* (21 August 1993): S1–14.

"That's Democracy." *The Economist* (26 June 1993): 43–44.

"Voting Against the Odds." *The Economist* (19 June 1993): 42.

The Washington Post (2 July 1993).

"Welding a Nation." *The Economist* (21 August 1993): A10–12.

Part IV

Conclusion

11

Positioning for the 1990s and Beyond

As we head through the last decade of the twentieth century and into the next, the international environment will be filled with pitfalls for the unwary as well as opportunities for those willing to do their proper due diligence. We believe that those countries that we have dubbed tigers exemplify the right package of factors: they have adopted export-oriented development models; a consensus within the leadership elite over policy direction; responsible and responsive legal systems, or at least are evolving in that direction; and some redistribution of national wealth. The others will continue to lumber along the elephant track, neither suffering economic disaster nor attaining developmental success. Some on the elephant track—most probably India—have the potential to shift on to the tiger path. Some on the tiger path—like Argentina and Kazakhstan—have the potential to fall back to the elephant track.

Implications of the Debate

The implications for Western policymakers and economic advisors are enormous. The field of economic development and reform is immensely more complicated and idiosyncratic than conventional wisdom would have us believe. The developed world must help create a hospitable environment for developing economies. This implies, most critically, progress on trade liberalization beyond the contentious GATT accord in 1993. However, at the same time we should be extremely wary of forcing dogmatic economic theories that emphasize abrupt, uniform, and far-ranging "reform and liberalization" on vastly differing developing economies. Ultimately the decision to undertake difficult economic reforms is the prerogative of the particular country. Although potential reformers may fear charges of imperialism or neocolonialism, they may find that these *shibboleths* are easily dismissed once the local population realizes benefits from economic liberalization.

The implications of the debate over political versus economic development are not always obvious for those conducting business abroad. For businesses, the most simple question is: How can I make a profit or avoid a loss in a world that develops according to these expectations outlined in this book? The risks range from opportunities lost to inaction, to massive losses arising from the wrong decision. For instance, the manager who chooses Nigeria as the site for his newest sales office may be in for some nasty shocks over the next decade, while the next person may regret their decision to avoid Argentina. At what point does a manager panic—pull out—or double the investment?

The stakes are at least as high for government policymakers. The United States has traditionally devoted itself (in between bouts of isolationism) to spreading the word of democracy and capitalism around the world. This has had mixed results. Most ominously, the decline of communism in Central and Eastern Europe has introduced democracy—of sorts—but has hardly brought the wonders of market capitalism to every country. Latin America shows more promise, but is still struggling to reconcile the demand of economic competition with growing social pressures. Africa's experience with both democracy and capitalism is at the lowest comparative level. The implosion of Rwanda in 1994 reflects some of the difficulties facing a continent where the very nature of the nation-state is under siege. The growing strength of Islamic radicalism, with its emphasis on meeting basic human needs in a number of communities in the Middle East, highlights the danger of ignoring poverty and misery indefinitely. Finally, the much-vaunted East Asian model is hardly one of Western liberal democracy, raising delicate issues that U.S. policymakers usually have carefully sidestepped, especially during the cold war. Thus, our expectations bring up a new set of questions for diplomats and government leaders. What is the role of aid in helping to support new democracies? What kinds of aid help and hinder development? Does food aid do more harm than good? What is the impact of military aid? How far should Western governments go in supporting their multinationals abroad? In the "new world disorder," do we ever revert to gunboat diplomacy?

The implications of the development debate, therefore, raise the tough decision about what type of world order we wish to foster. While a single, major war appears unlikely, there is a strong possibility that the next twenty years will be one of the most violent periods in centuries. The nation-state is under intense pressure: the Soviet Union and Yugoslavia have split into a number of new states, while Liberia and Somalia are reminiscent of the nursery rhyme's Humpty Dumpty, who could never be put together again. From a U.S. perspective, is it worth the resources to attempt to reconstruct these countries?

Considerations for Business

Some of our assumptions indicate that governments and businesses will require answers more and more urgently over the balance of the 1990s and beyond. Both sectors are driven by the need to strategize. Governments are pushed by national interests that include employment, competition in foreign markets, and national security. The private sector is faced with a delicate balance between risk and opportunity that will test its ingenuity. Our expectation is that enormous opportunities will be available in certain marketizing economies, ranging from Slovenia and Vietnam to Argentina and China. Other countries not covered in this book, but worth consideration as tigers, include Tunisia, Indonesia, Cyprus, Uruguay, and Myanmar (Burma).[1] However, these opportunities are associated with high risks as well. In addition to the usual risks of political turmoil, expropriation, currency instability and inconvertibility, and so on, there is a new category that we will call "bandwagon risk." The trouble is that everyone is anxiously awaiting the next tiger, and is ready to leap on the bandwagon well before a tiger has defined itself. This was illustrated by the run up in stock prices in various high-risk countries of Latin America in the late 1980s—no one wanted to miss the next Mexico or Chile. The best time to enter a market is before everyone else has decided to rush in too. On the other hand, as *The Wall Street Journal* points out, "[T]here's such a thing as being too early in jumping on an investment trend."[2] Ideally, the hardy investor should be seeking low-cost, high-risk opportunities. Ill-timed investor behavior can also have a negative impact on a country's development—too much capital too fast can result in inflation and currency problems.

For those who do make the right decisions (or educated guesses), the stakes are tremendous. According to *The Emerging Markets Analyst*, marketizing or "emerging" markets represented less than 5 percent of the world's total stock market capitalization at the end of 1991, but their performance and growth far outstripped that of developed capital markets. According to these calculations, a $100 index-linked investment made in 1975 in the best performing G7 (Canada, France, Germany, Italy, Japan, United Kingdom, and the United States) stock market would have yielded $21,633 by the end of 1991, equivalent to an annual average compound return of 40 percent. The best performing emerging markets over the same period would have yielded a staggering $552,354,010—an annual average compound return of 164 percent. The emerging markets have continued to expand since 1981 and are perceived as one of the hot markets in the 1990s, despite what will probably be the temporary bear markets of 1994–1995.

Are investors considering the wide range of developing countries? The secondary market for Emerging Market country debt, where these liabilities are sold at a discount, has expanded from an aggregate volume of $1 billion in 1984 to an estimated $600 billion in 1992, transacted among market professionals and institutional and retail investors around the world.[3] Add to this newly issued bonds: in the 1989–92 period, there were over 200 Latin American voluntary "new issue" bond transactions totalling over $20 billion in nine currencies for sovereign, sovereign-supported, and corporate borrowers from six countries.[4] This does not take into consideration money raised in bond markets by other countries, such as China, India, the Philippines, and Indonesia, or the creation of country funds that trade on the New York Stock Exchange. Although Mexico's problems in 1994 and 1995 set the market back to a "submerging market," the potential of a strong renewal is probable later in the decade.

Marketizing economies have vast infrastructure needs. For example, Japan's Long-Term Credit Bank thinks that Asia, excluding Japan, will spend $1 trillion on infrastructure over the next decade.[5] This can offer excellent opportunities for investors in international bond and equity issues. Usually state-owned or parastatal utility companies (such as Korea Electric Power Company, Malaysia's Tenaga Nasional Berhad, or Chile's Endesa) offer excellent returns on investment and are sought by investors. Other companies, like Chile's CTC (the privatized telephone company), have become highly popular on major stock markets.

Portfolio investment is hardly the only opportunity for private sector companies anxious to get in on the birth pangs of an emerging tiger.

Foreign direct investment is often the next step. In this regard, the infrastructure defects of tiger cubs require special attention. Physical infrastructure—transport, telecommunications, roads, electricity—often is a binding constraint for companies looking to invest in young developing markets. In certain cases, such as Peru and Nigeria, physical facilities have deteriorated so badly that this now constitutes a bottleneck to development plans.

It is not a coincidence that some of the hottest investments in developing countries are infrastructure linked. One opportunity that has already surfaced is the Emerging Markets Telecommunications Fund, which invests in phone companies of developing countries, mostly Latin America.[6] (Anyone who has ever tried to place a conference call from Boston to Brasilia or even within the city of Lagos will appreciate the need for such an investment.) Investors do not want to miss an issue that could duplicate the smash-hit 1991 offering by Telefonos de Mexico.

Another example of this trend is the early success of a small Canadian-based trading company, MTC Electronic. This company plans to become the biggest provider of cellular mobile phone service to China. MTC views China as a huge opportunity, given its enormous market and the inadequate condition of existing telephone service.[7] However, the risks faced by MTC highlight those confronting all investors in this market. For one thing, the company relies on its ability to sell stereos and television sets exported from China to convert into hard currency its revenue from sales in China. If this market dries up, so does MTC's ability to repatriate profits.

Another concern for telecommunications and utilities investors is the propensity of some developing countries to institute price controls on necessities such as electricity and water. Many infrastructure-linked activities have little potential for high margins, partly due to regulatory burdens. In addition to repatriation problems, some countries still maintain nationalistic provisions on many infrastructure industries. Even Mexico, for instance, still has a constitution that reserves ownership and distribution of many public services to the state.

Thus, the infrastructure market presents broad risks as well as opportunities for the stalwart investor. It is not surprising that many multinationals instead look to natural resource extraction as the primary activity on entering (or reentering) an emerging economy. In the past few years, for instance, large international mining companies have increased their operations in Latin America, drawn by the region's economic reforms and scale of untapped mineral resources. Latin American countries often offer the mining industry preferred tax treatment, security of mineral tenure, government joint ventures, prompt government assistance and response to development plans, and guaranteed repatriation of profits and even capital.

Latin America indeed offers mining companies a wealth of skilled labor at relatively competitive wages, as well as faster environmental and safety approvals. One warning is provided by the example of Brazil. In contrast to its neighbors, Brazil inserted a clause into its 1988 constitution that prohibits foreign companies from holding a majority interest in mining operations; this has led to a precipitous plunge in investment by international mining companies. While interest in foreign investors is at an all-time high regionwide, the Brazilian constitution serves as an instructive reminder that foreign exploitation of natural resources remains a delicate issue. The other warning for this sector relates to one of our original assumptions: relative commodity prices will be depressed over the next decade as technological advances gradually dampen demand for raw material inputs.

Nonetheless, these two sectors—infrastructure and natural resources—will probably attract the most attention from foreign investors during the early stages of the upcoming development era for our emerging tigers. This includes Kazahkstan's and Argentina's hydrocarbon reserves, China's and Vietnam's vast infrastructural needs, and possibly Morocco's phosphates. Companies that choose to jump in (or hang back) will have to be especially nimble because the volatility of the environment suggests that radical changes midstream are possible. This implies that companies will have to be alert to early warning signs of a shift, in both tigers and elephants. Some early signs may be clearly evident, even dramatic, such as the end of the U.S. embargo on Vietnam in 1994. Others can be quite subtle, such as leadership nuances in China or Argentina. The high likelihood of major shifts, along with the bandwagon risk described earlier, suggests that companies will have to be prepared for quick movement to take advantage of, or to protect themselves from, changes that can transform an elephant into a tiger, or vice versa.

Consumer goods represent one more area that investors should consider in the marketizing economies discussed in this book. Guangdong Province in China is one of the largest markets for shampoo for Proctor and Gamble. As the population of the various countries that hit the tiger track move up the socioeconomic ladder and the size of local middle classes grow, options for consumer product market expansion must be considered. Countries such as Argentina, India, and Indonesia have sizeable middle classes. It is probable that as Vietnam advances, its middle class will demand more consumer goods; marketing firms are already analyzing local conditions. In fact, a study was conducted in October, 1993 that indicated that Vietnam offered potential for such goods as blue jeans and athletic footgear, partly because one-fifth of the population was under twenty years of age.[8] Businesses, however, must clearly consider all their political risks as well as their social impact—crass consumerism can result in a cultural backlash that will negate hard-earned market share.

Tough Policy Choices

Policymakers in the public sector face a more basic dilemma in positioning themselves for the 1990s based on our expectations. Our sequencing assumption posits a basic tension between democracy and economic progress that has long troubled Western leaders. We believe that these tensions will grow rather than diminish over the forecast period, fed by ethnic/racial tensions that magnify the misery of poverty and income distribution inequities.

Without the cold war's "chill factor" that froze many regional and ethnic tensions in place, the views of Alexis de Tocqueville that the association of economic development, particularly at a rapid pace, with political instability may be as valid at the end of the twentieth century as they were at the time of the French Revolution. According to Tocqueville, the French Revolution was preceded by "an advance as rapid as it was unprecedented in the prosperity of the nation."[9] This implies that even countries (such as China and Vietnam) that desire to be soft authoritarians face tough challenges in the future—a factor that policymakers should consider carefully. Even in India, where the track record on economic reform has historically been less dynamic than in other countries, observers saw signs that economic advances had political repercussions as early as the 1960s: India's "economic development, far from enhancing political stability, has tended to be politically unstabilizing."[10]

As a result, the West will confront a growing threat of dramatic social and political upheaval in the developing world radicalism in the coming decades. Nourished by poverty and sometimes linked to Islamic radicalism, hostility to the West will grow as developing countries make limping advances toward economic reform, but are often stymied by Western protectionism, recession, and domestic political imperatives. In other cases, boom-and-bust cycles may aggravate political tensions. The danger is that as structural adjustment takes place, poverty rises and the gap between rich and poor widens at first. Moreover, investment is unlikely to arrive simultaneously in all countries that are opening up their economies to attract foreign capital.

From the governmental point of view, our sequencing assumptions present a distasteful conundrum. On the one hand, democracy in developing countries can be seen as messy, annoying, even unnecessary to the economic reform process. *The Wall Street Journal,* in an article aptly titled, "Why Global Investors Bet on Autocrats, Not Democrats," used the contrasting examples of India (democratic, but unsuccessful in economic development) and China (economically successful, but closed politically) to demonstrate the advantages of political order and stability. The *Journal* further suggests that the "dictatorship" model has been generally accepted in Asia, but "the real question for Asia is whether the benign dictatorship is a stable model," suggesting that a difficult transition to democracy may lie ahead.[11] It should be remembered that development has also posed problems for authoritarian as well as democratic governments as long ago as the French Revolution and the Porforian regime in pre-revolutionary Mexico.

We would argue, of course, that once economic reform and prosperity have been achieved, political reform is much easier than if things are done the other way around. Herein lies the quandary. Our model might suggest that Western policymakers encourage the economic reform process in the developing world while turning a blind eye to benign dictatorships (as they did, in fact, for the Asian tigers). Specifically, the West should invest in China rather than punish it, politely deplore Fujimori's autogolpe while encouraging companies to trade and invest in Peru, and even hope for the emergence of an enlightened strongman in Argentina and Brazil. Given the western commitment to human rights and democracy in its foreign policymaking, this is not a possibility. It may also not be moral.

However, the stakes for Western governments could not be higher. From an economic and financial point of view, it is essential that the developing world become reliable and efficient partners in the world trade and investment game. Huge markets in China and South Africa, for instance, are virtually untapped and could prove to be worldwide engines of growth under the right circumstances. The political stakes are even higher. The threat of disunity, of fragmentation as in Somalia or Yugoslavia, has never been greater. Important portions of the developing world threaten to break up into tribal enclaves, economically inefficient and politically unreliable. We need only look to Bosnia to observe the emasculation of economic hopes by political tribalism. Other countries, armed with weapons of mass destruction and centuries of suspicion of the West, could become formidable enemies. The governments, populations, and businesses of industrialized nations will be better served by developing nations that share values of open economic and political systems. As veteran Indian journalist Pranay Gupte noted in his biography of Indira Gandhi: "The most immediate—and fundamental—contribution that Washington and its Western allies could make in the region, however, is through the promotion of democracy.... The authoritarian caudillo is an attractive bet for the superpowers, but these dictators are really short-term political animals—they are sprinters rather than long-distance runners who could sustain the long-term dynamic of development."[12]

Thus, the decisions made by Western policymakers and businessmen on dealing with developing countries over the next decade will help to shape a new world. The emergence of new tigers, and the plodding progress of old elephants, will present unforeseen challenges and opportunities to decision makers. Considering these factors, this also means that policymakers will have to consider new actors on the global stage. As Irving Louis Horowitz notes, "It may well be that the next century will witness new players of world historic importance—among them China, Canada, Brazil, Australia and South Africa, for starters."[13] We are in a position to help or hinder world development, profit or lose through participation in emerging markets, and alleviate or magnify the misery of millions in the process.

Final Words

In conclusion, we regard the development prospects of Argentina, China, Kazahkstan, Morocco, Slovenia, and Vietnam as high. Our new tiger selections have considerable potential to join an older tiger club of

Chile, Czech Republic, Hong Kong, Malaysia, Mexico (albeit in a weakened position after December 1994), Singapore, and Thailand.

At the same time, we also believe that Brazil, India, Nigeria, Peru, and Venezuela will continue to lumber along the elephant track. Of this group of elephants, India has the best chance of finding a path to more sustained and diversified growth.

For those looking for business opportunities, the play is on infrastructure, mining, direct investment in manufacturing, and certainly retailing to selected markets. For businessmen playing the market for sovereign or corporate bond issues from the developing countries, the trick is to be increasingly selective and into the balance of the 1990s and into the 2000s.

Government policymakers face a different, yet related set of issues. The new tigers offer a challenge of being largely transitional in both their political and economic development. A shared set of core values—favoring open political and economic systems—is more supportive of maintaining a global capitalist system in the long run than the emergence of anti-Western and xenophobic regimes, bent on disrupting the evolution of an interdependent world order.[14] Support for capitalist development strategies of course reinforces the mutual flow of investment and commerce.

A new round of the development game is now afoot in the aftermath of the cold war—a game involving high risks and high rewards. The players in this global contest will find that the rules have changed in the 1990s. Gone are the old cold war parameters that froze traditional conflicts, confined policy directions, and forced governments to choose up sides between East and West. The new cleavage is between the proponents and the opponents of international integration. We believe that the proponents of globalization will be the winners, and our new tigers will be among those counties leading the way and benefitting from the results.

Notes

1. See Amy Kaslow and Scott B. MacDonald, "Coming Into Its Own: Tunisia's Economy," *Middle East Insight IX,* 6 (September-October 1993), 40-421, and Mya Than and Joseph L. H. Tan, *Myanmar Dilemmas and Options: The Challenge of Economic Transition in the 1990s* (Singapore: Institute of Southeast Asian Studies, 1990).
2. William Power, "Fund That Invests in Third-World Phone Firms Lures Investors Despite Companies' Hang-Ups," *The Wall Street Journal* (13 July 1992).
3. Suhas Ketkar and Stefano Notella, *An Introduction to Emerging Countries Fixed-Income Investments* (New York: CS First Boston, May 1993), 28.

4. Ibid.
5. "Infrastructure in Asia: the Trillion-Dollar Dream," *The Economist* (26 February 1994): 66.
6. Power, "Fund that Invests in Third-World Phone Firms Lures Investors Despite Companies' Hang-Ups."
7. Anne McGee, "Stock of Little MTC Electronic Zooms on China Hopes," *The Wall Street Journal* (9 December 1992).
8. Kevin Goldman, "U.S. Ad Firms Set Sights on Reaching Vietnamese," *Asian Wall Street Journal* (6 February 1994): 6.
9. Alexis de Tocqueville, *The Old Regime and the French Revolution* (Garden City, N.Y.: Doubleday, 1955), 173.
10. Bert Hoselitz and Myron Weiner, "Economic Development and Political Stability in India," *Dissent 8* (Spring 1961): 173.
11. Michael Sesit and Robert Steiner, "Why Global Investors Bet on Autocrats, not Democrats," *Wall Street Journal* (12 January 1993).
12. Pranay Gupte, *Mother India: A Political Biography of Indira Gandhi* (New York: Charles Scribner's Sons, 1992), 420.
13. Irving Louis Horowitz, "Histories, Futures, and Manifest Destiny," *Society 31*, 5 (July/August 1994): 12.
14. See Larry Diamond, "The Global Imperative: Building a Democratic World Order," *Current History* (January 1994): 2.

References

Abrams, Elliott. "How To Avoid the Return of Latin Populism." *The Wall Street Journal* (21 May 1993).

Diamond, Larry. "The Global Imperative: Building a Democratic World Order." *Current History* (January 1994).

Emerging Markets Analyst, 1992.

Fidler, Stephen. "Backward Step in Latin America." *Financial Times* (28 May 1993).

Goldman, Kevin. "U.S. Ad Firms Set Sights on Reaching Vietnamese." *Asian Wall Street Journal* (6 February 1994).

Gupte, Pranay. *Mother India: A Political Biography of Indira Gandhi.* New York: Charles Scribner's Sons, 1992.

Horowitz, Irving Louis. "Histories, Futures, and Manifest Destiny." *Society 31*, 5 (July/August 1994).

Hoselitz, Bert and Myron Weiner. "Economic Development and Political Stability in India." *Dissent 8* (Spring 1961).

"Infrastructure in Asia: the Trillion-Dollar Dream." *The Economist* (26 February 1994).

Kaslow, Amy and Scott B. MacDonald. "Coming Into Its Own: Tunisia's Economy." *Middle East Insight IX*, 6 (September-October 1993).

Ketkar, Suhas and Stefano Notella. *An Introduction to Emerging Countries Fixed-Income Investments.* New York: CS First Boston (May 1993).

McGee, Anne. "Stock of Little MTC Electronics Zooms on China Hopes." *Wall Street Journal* (9 December 1992).

"Perilous Moment" (Editorial). *Financial Times* (7 December 1992).

Power, William. "Fund that Invests in Third-World Phone Firms Lures Investors Despite Companies' Hang-Ups." *Wall Street Journal* (13 July 1992).

Prowse, Michael. "Miracles Beyond the Free Market." *Financial Times* (26 April 1993).

Sesit, Michael and Robert Steiner. "Why Global Investors Bet on Autocrats, not Democrats." *Wall Street Journal* (12 January 1993).

Than, Mya and Joseph L. H. Tan. *Myanmar Dilemmas and Options: The Challenge of Economic Transition in the 1990s.* Singapore: Institute of Southeast Asian Studies, 1990.

de Tocqueville, Alexis. *The Old Regime and the French Revolution.* Garden City, N.Y.: Doubleday, 1955.

Index